STATE MACHINES
IN
VHDL
Trigonometric Functions
Vol. 4

State Machine Design for
Arithmetic Processes

Daryl Ray Hawkins

Revision Dates
June 2016 First Release

Table of Contents

Overview

Trigonometry is a highly involved subject with many applications and uses. Some review is provided to the reader, but only relevant information for understanding and designing hardware trigonometric functions.

Functions covered in this book are those that are useful to electrical engineers and computer science professionals: sine, cosine, and tangent, and their reciprocals: cosecant, secant, and cotangent. Inverse functions: arcsine, arccosine, and arctangent, as well as their reciprocals: arccosecant, arcsecant, and arccotangent.

Two architectures are covered: Table-Based and Real-Time solutions. Table-based solutions are provided for all twelve functions in both degrees and radians, totaling twenty-four designs. Real-Time solutions include: Taylor-Series utilizing polynomials and Borchardt-Gauss using iterated means (Arithmetic-Geometric Mean).

Taylor-Series implementations are for the primary core group sine, cosine, and tangent. Borchardt-Gauss iteration target their inverse functions: arcsine, arccosine, and arctangent. Six designs in all, radians only.

All implementations are fixed-point. Real-Time implementations can use any of the multiplier, divider, and square root modules detailed in Volumes 2,3, and 5 of this books series. Hence, performance can be scaled by the capability of the modules selected.

Copies of all source code, in text file form, used in this book may be purchased through the web site listed below -- under publications, or pubs.

http://www.hawkinseng.com

1 Prerequisites

State machine techniques used throughout this book are covered in the "STATE MACHINES IN VHDL *Composition* Vol. 1" book, which is a prerequisite, and should be reviewed by the reader.

Additionally, "STATE MACHINES IN VHDL *Multipliers* Vol. 2" and "STATE MACHINES IN VHDL *Dividers* Vol. 3" contain useful information on formatting and rounding techniques. These subjects will not be repeated here.

Finally, reviewing the IEEE 754-1985/2008 specifications is also recommended.

2 Back to School

Even though this book is not a textbook on trigonometry nor is it intended to teach trigonometry, reviewing the basics is necessary for the reader to follow terms and concepts that are used in later chapters. Figure 1 provides graphic illustrations that are common in trigonometry.

- Theta (θ) is the angle between the adjacent and hypotenuse.

- The base results of a function involve a right triangle in a single quadrant, like 0 to 90°. Core functions are extended to all quadrants (multi-valued), but inverse functions have restricted domains, occupying the equivalent of only two quadrants.

- A unit circle has a radius of 1

- A circle may be defined in degrees (°) or radians (rad)

- Some functions exhibit asymptotic behavior as they approach quadrant boundaries. Function values near those boundaries approach infinity and cannot represented. Values on those boundaries cannot be defined.

- The beginning or end of each quadrant can be represented in either degrees or radians.

0°/0° quadrant I	or	0 rad	= 0
90°/+90° quadrant I	or	1.5708 rad	= $\pi/2$
180°/0° quadrant II	or	3.1416 rad	= π
270°/-90° quadrant III	or	4.7124 rad	= $3\pi/2$
360°/0° quadrant IV	or	6.2832 rad	= 2π

In the descriptions that follow, each subsection provides the function's nomenclature, domains, as well as a graphical representation. The latter illustrated in degrees for simplicity.

Note: *Since the range of numbers that can be represented depends on the size of the fixed-point format chosen in the implementation, resulting values equal to or greater than the maximum value will equal that maximum value, positive or negative. This is sometimes referred to as saturation. Even though terms like +∞ or -∞ are used, the maximum value can only be within the range of the number format. An error flag is asserted when boundaries are exceeded.*

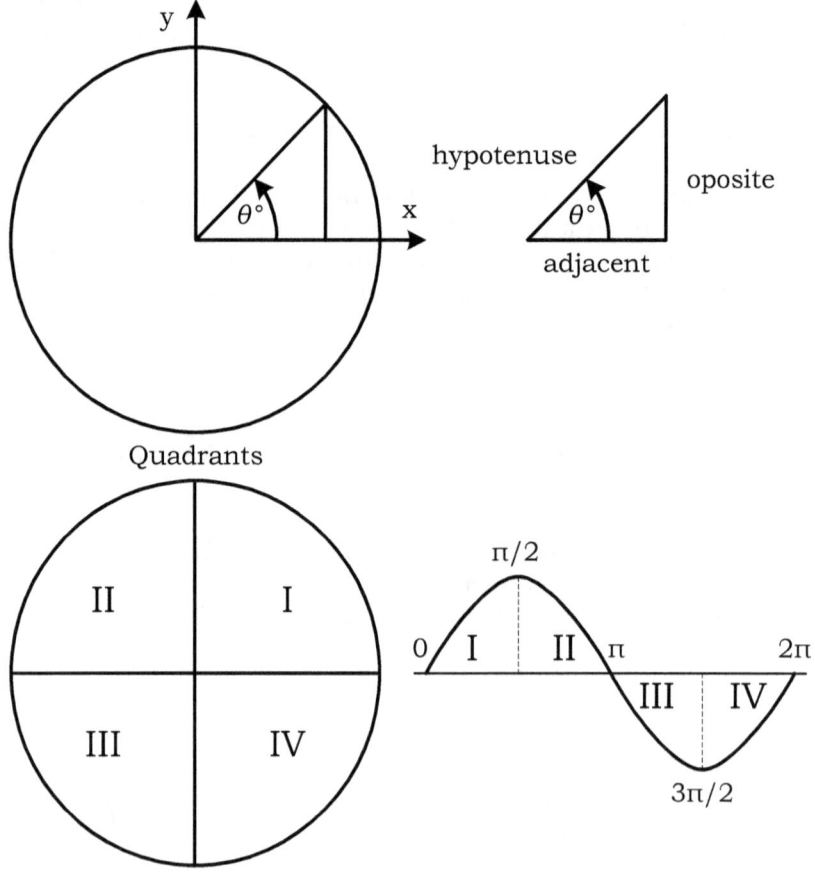

Figure 1

2.1   Core Functions and Their Reciprocals

Sine θ = ratio of opposite/hypotenuse sides, which is also referred to as the rise of the opposite segment.

sin($θ$), returns a value between -1 to +1 on any angle throughout the circle. Quadrant sequence is:

- 0 to 90° (π/2) returns 0 to +1
- 90° to 180° (π) returns +1 to 0
- 180° to 270° (3π/2) returns 0 to -1
- 270° to 360° (2π) returns -1 to 0

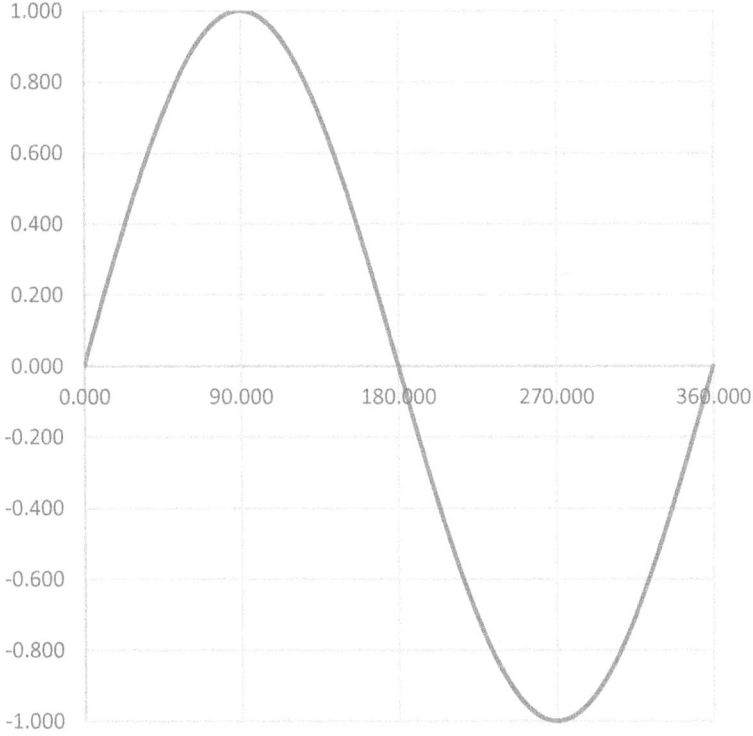

Cosine *θ* = ratio of the adjacent/hypotenuse sides, which is also referred to as the run of the adjacent segment.

cos(*θ*), returns a value between -1 to +1 on any angle throughout the circle. Quadrant sequence is:

- 0 to 90° (π/2) returns +1 to 0
- 90° to 180° (π) returns 0 to -1
- 180° to 270° (3π/2) returns -1 to 0
- 270° to 360° (2π) returns 0 to +1

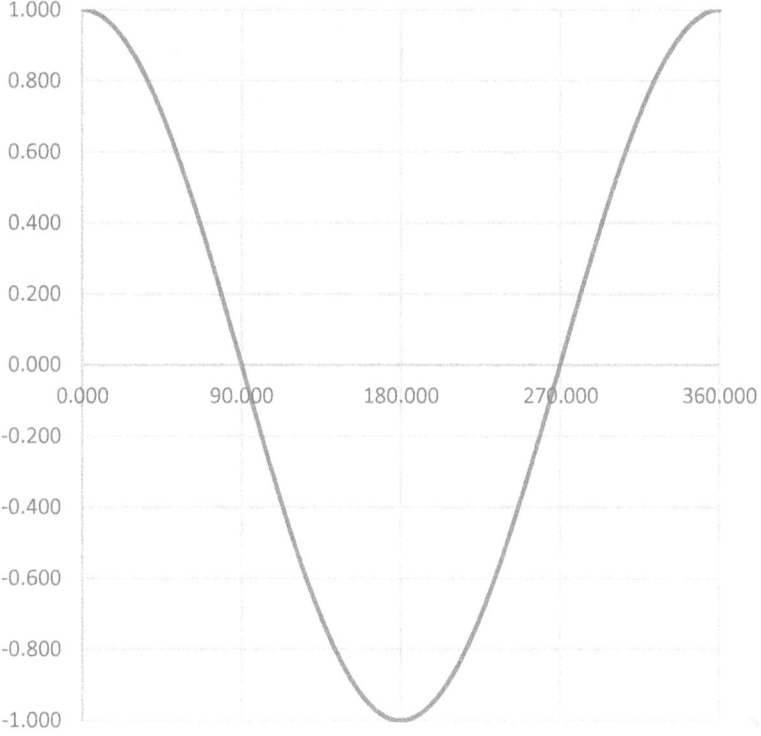

Tangent θ = ratio of the opposite/adjacent, which is also the slope of the hypotenuse segment.

tan(θ), returns a value between 0 and +/-∞ on any angle throughout the circle. Quadrant sequence is:

- 0 to just less than 90° ($\pi/2$) returns 0 to +∞
- Just more then 90° to 180° (π) returns -∞ to 0
- 180° to just less than 270° ($3\pi/2$) returns 0 to +∞
- Just more then 270° to 360° (2π) returns -∞ to 0

Note: *A tangent of absolute 90° or 270° has no solution because the slope (function of the angle) is a vertical line, ∞, and is neither + or -. The immediate switch from +∞ to -∞, and vice-versa, when crossing the 90° and 270° boundaries has to do with how we define a slope. Slopes are evaluated from left to right. If the slope points down it is a negative slope; whereas if it points up it is a positive slope.*

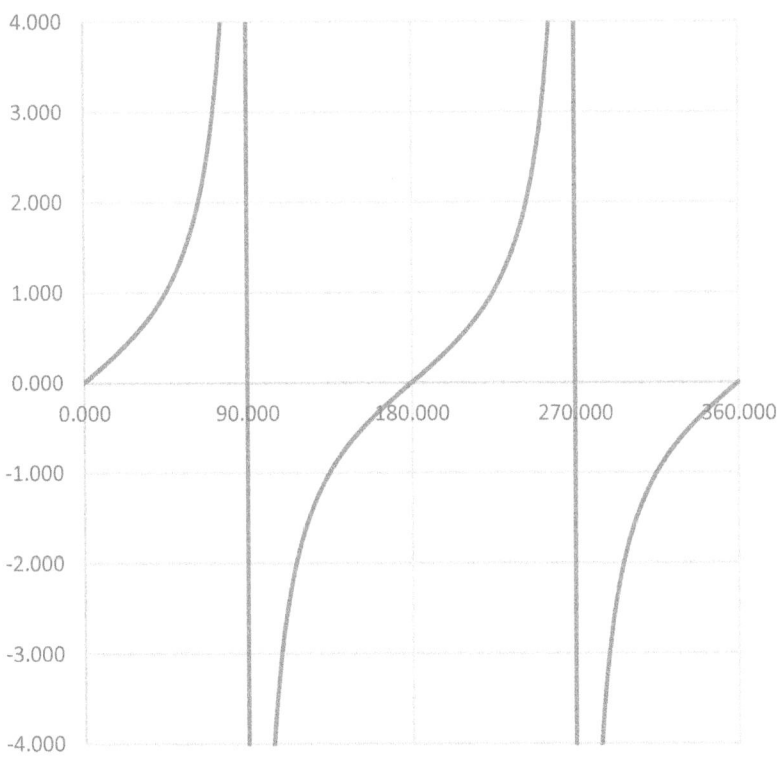

Cosecant θ = ratio of hypotenuse/opposite, which is the reciprocal of the sine function, $1/\sin(\theta)$.

$\csc(\theta)$, returns a value between 1 and +∞, or between -1 and -∞. Quadrant sequence is:

- Just more than 0 to 90° ($\pi/2$) returns +∞ to +1
- 90° to just less than 180° (π) returns +1 to +∞
- Just more than 180° to 270° ($3\pi/2$) returns -∞ to -1
- 270° to just less than 360° (2π) returns -1 to -∞

Note: *A cosecant of absolute 0 and 180° has no solution because the value is ∞ with no distinction between + or -.*

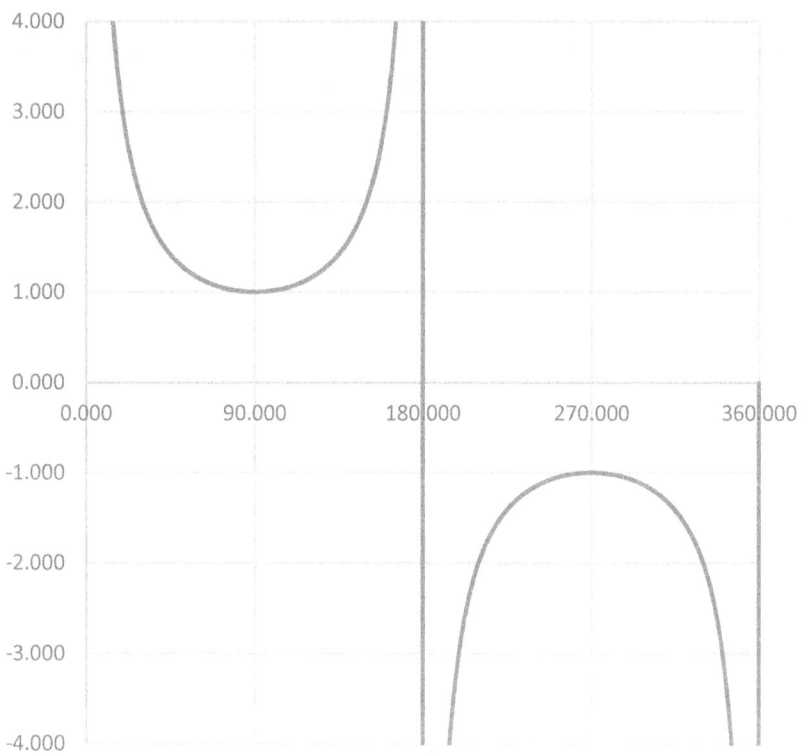

Secant θ = ratio of hypotenuse/adjacent, which is the reciprocal of the cosine function, $1/\cos(\theta)$.

$\sec(\theta)$, returns a value between 1 and $+\infty$, or between -1 to $-\infty$. Quadrant sequence is:

- 0 to just less than 90° ($\pi/2$) returns +1 to $+\infty$
- Just more than 90° to 180° (π) returns $-\infty$ to -1
- 180° to just less than 270° ($3\pi/2$) returns -1 to $-\infty$
- Just more than 270° to 360° (2π) returns $+\infty$ to +1

Note: *A secant of absolute 90° and 270° has no solution because the value is ∞ with no distinction between + or -.*

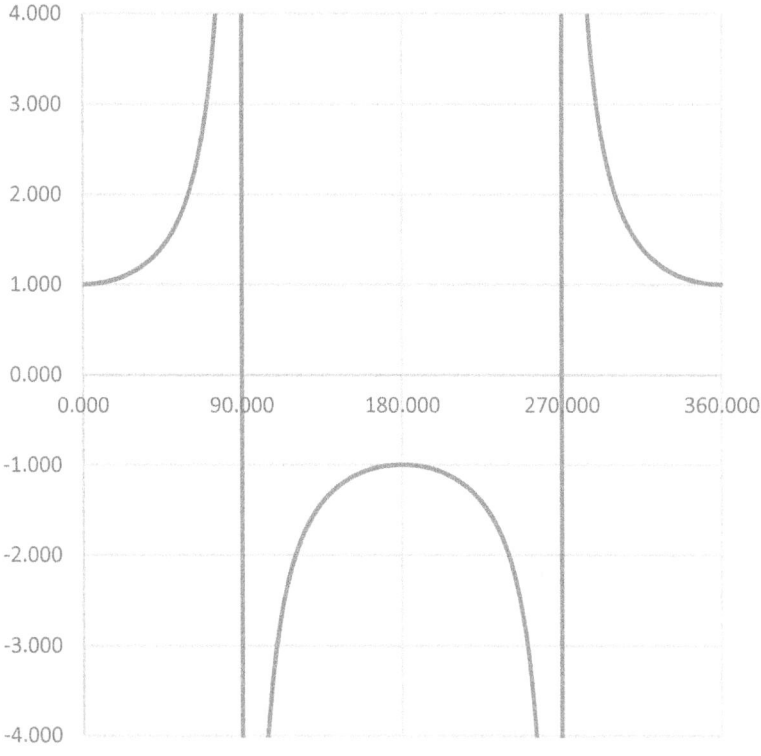

Cotangent *θ* = ratio of adjacent/opposite, which is the reciprocal of the tangent function, 1/tan(*θ*). As with tangent, it repeats itself every 180°.

cot(*θ*), returns a value between 0 and +/-∞. Quadrant sequence is:

- Just more than 0 to 90° (π/2) returns +∞ to 0
- 90° to just less than 180° (π) returns 0 to -∞
- Just more than 180° to 270° (3π/2) returns +∞ to 0
- 270° to just less than 360° (2π) returns 0 to -∞

Note: *A cotangent of absolute 0 and 180° has no solution because the value is ∞ with no distinction between + or -.*

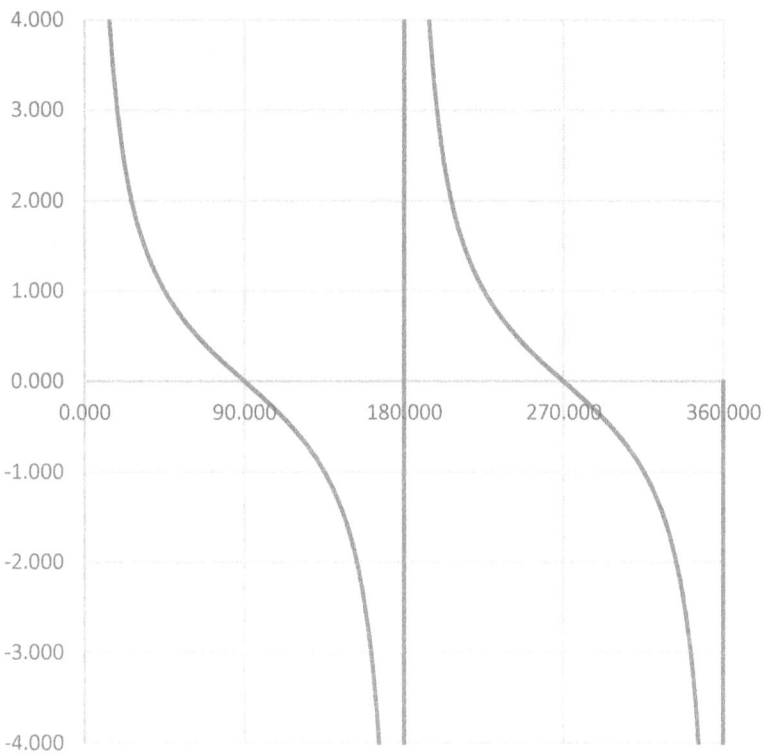

2.2 Inverse Functions and Their Reciprocals

Inverse or arc functions are those that accept the rise (sine), run (cosine), or slope (tangent) value and return the corresponding angle. Unlike the core functions where the input can span multiple quadrants by way of degrees or radians, inverse functions cannot. Conventional inverse functions operate within a restricted domain or subset of two quadrants, evaluating a single input called the principle value and returning a limited range -90 to +90°, or 0 to +180°. Graphical representation in some functions are non-continuous or broken.

Inverse Sine x = input is the rise from -1 to +1.

arcsin(x) , returns an angle between -90 and +90°. Domain sequence is:

- -1 to 0 returns -90° (-π/2) to 0
- 0 to +1 returns 0 to +90° (π/2)

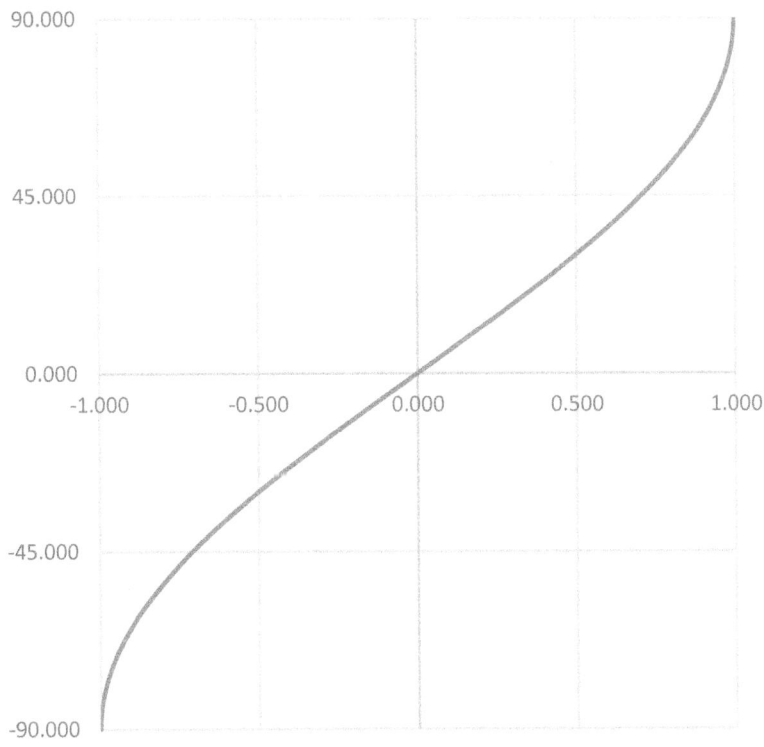

Inverse Cosine x = input is the run from +1 to -1.

arccos(*x*), returns an angle between 0 and +180°. Domain sequence is:

- +1 to 0 returns 0 to +90° (π/2)
- 0 to -1 returns +90° (π/2) to 180° (π)

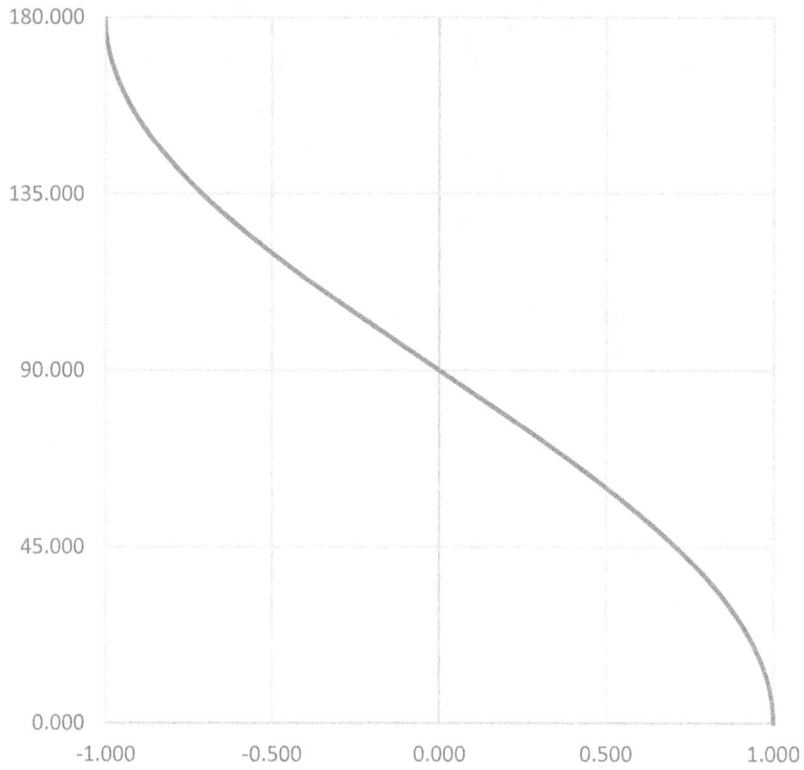

Inverse Tangent x = input is the slope from -∞ to +∞.

arctan(x), returns an angle between just more than -90° and just less than +90°. Domain sequence is:

- -∞ to 0 returns just more than -90° (-π/2) to 0
- 0 to +∞ returns 0 to just less than +90° (π/2)

Note: *Just as with tangent, an absolute angle of -90° (270°) or +90° has no solution. Likewise, inputs of +/- ∞ have no representation.*

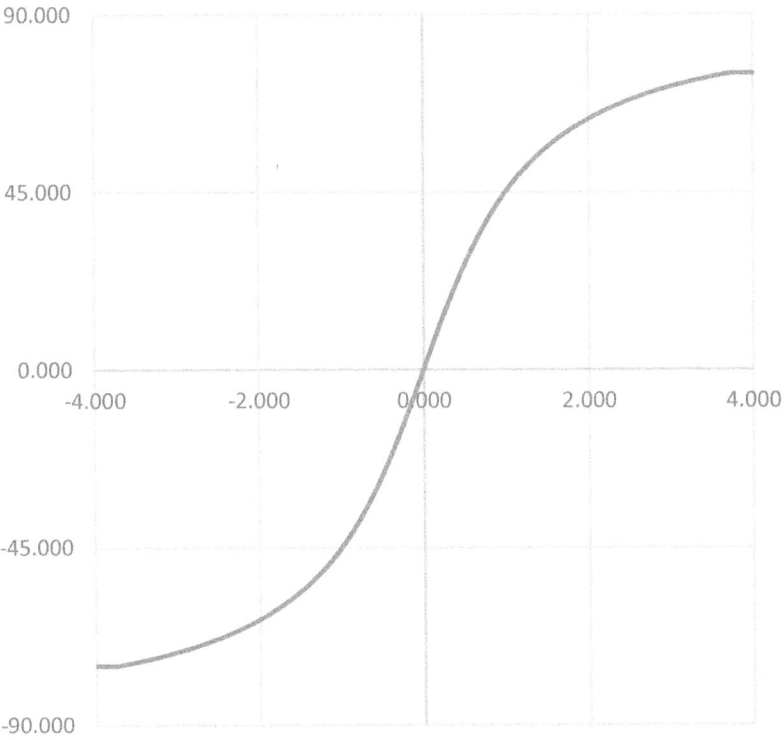

Note: *The x axis boundaries of +/- 4 are the same as the input range of table-based inverse functions, actual limits are +/- ∞.*

Inverse Cosecant x = input is between -∞ to -1 and +1 to +∞. It is the reciprocal of the Inverse Sine function, arcsin(1/x).

arccsc(x), returns an angle between -90° and just less than 0, or just more than 0 to +90°. Domain sequence is:

- -1 to -∞ returns -90° (-π/2) to just less than 0
                 -- OR --
- +1 to +∞ returns 90° (π/2) to just more than 0

Note: *Just as with Cosecant, an absolute angle of 0 has no solution. Likewise, inputs of +/- ∞ have no representation.*

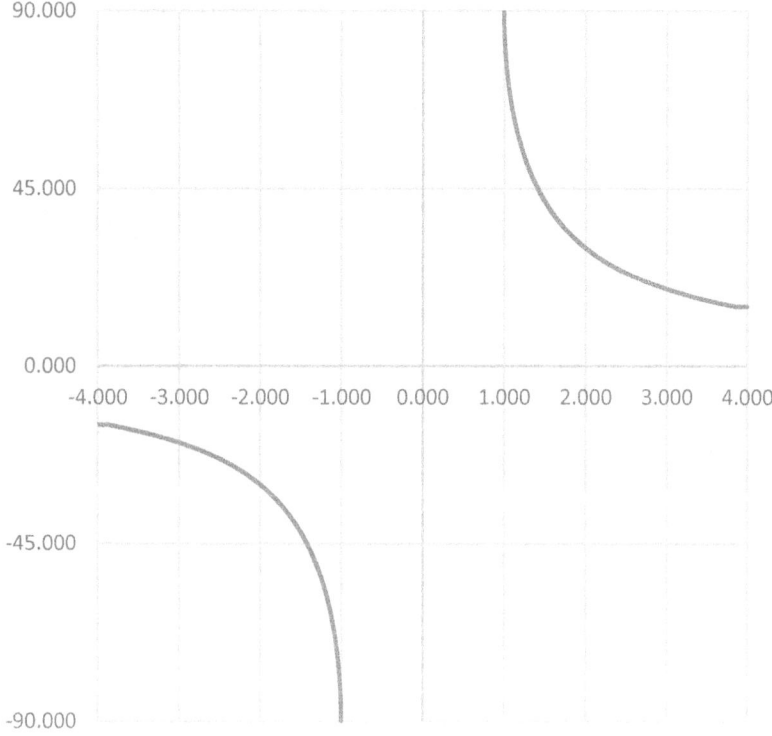

Note: *The x axis boundaries of +/- 4 are the same as the input range of table-based inverse functions, actual limits are +/- ∞.*

Inverse Secant x = input is between -∞ to -1 and +1 to +∞, it is the reciprocal of the Inverse Cosine function, arcos(1/x).

arcsec(x), returns an angle between just more than 90° to 180° or 0 to just less than 90°. Domain sequence is:

- -∞ to -1 returns just more than 90° (π/2) to 180° (π)
  -- OR --
- +1 to +∞ returns 0 to just less than 90° (π/2)

Note: *Just like Secant, an absolute angle of 90° and -90°(270°) has no solution. Likewise, inputs of +/- ∞ have no representation.*

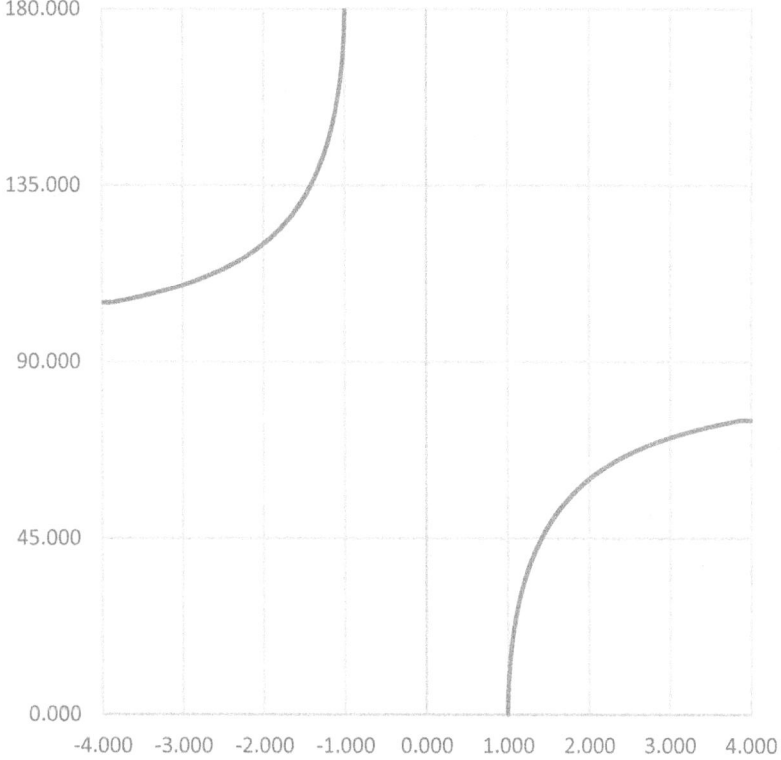

Note: *The x axis boundaries of +/- 4 are the same as the input range of table-based inverse functions, actual limits are +/- ∞.*

Inverse Cotangent *x* = input is between -∞ to +∞, is the reciprocal of the Inverse Tangent function, arctan(1/*x*). An alternative is π/2 – tan(*x*), which is sometime preferred.

arccot(*x*), returns an angle just less than 180° and just more than 0. Domain sequence is:

- -∞ to 0 returns just less than 180° (π) to 90° (π/2)
- 0 to +∞ returns 90° to just more than 0

Note: *Just like Secant, an absolute angle of 0 and 180° has no solution. Likewise, inputs of +/- ∞ have no representation.*

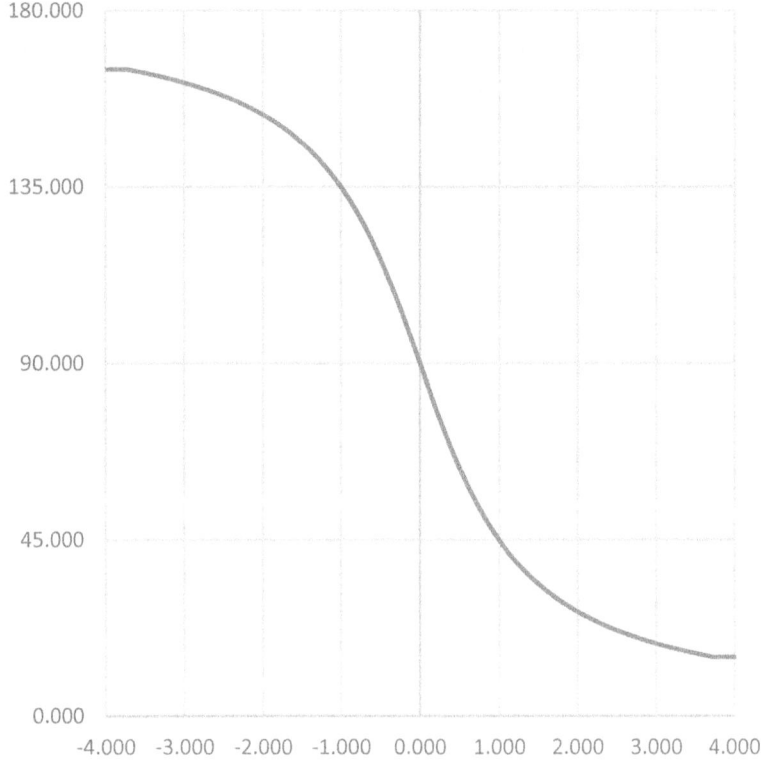

Note: *The x axis boundaries of +/- 4 are the same as the input range of table-based inverse functions, actual limits are +/- ∞.*

3 Table-Based Solutions

Table based solutions have been around the longest, utilizing pre-computed values and indexing them by either degrees or radians, or the inverse in arc functions.

IEEE has a standard package with the same core and inverse trigonometric functions covered in this book -- MATH_REAL.VHD. These functions can be used during instantiation to initialize block RAM, or generate text files used to load external memory. The point being, the computational work has already been done.

Then why use a state machine?

There are boundaries that require checking, exceptions that must be handled, errors reported, conversions made, and so on. Having a self-contained module that accepts a parameter then returns the result, flagging errors if applicable, is what we want. Keeping all associated overhead within the module greatly simplifies a high-level design. This of course is not without difficulty.

Properly scaling memory based on operand size (fixed-point only), the type of adders used in boundary and error checks, and sign conversions are just a few. As in the design of multipliers and dividers, the use and type of adders will have the greatest impact on performance. Additionally, there may be certain behaviors within the package functions that require special handling in some user defined way.

As in *Chapter 2*, this chapter likewise is divided into to two sections, one for core functions and their associated reciprocals, and the other for the inverse functions and their reciprocals. Partitioning in this fashion simplifies subject coverage and implementation of the source code, in particular, the package files. Core functions are the most commonly used, accepting an angle in either degrees or radians and outputting the function of that angle. Conversely, inverse functions accept the function of that angle and outputs an angle, in either degrees or radians.

Sizing Memory

Properly sizing memory, which is a limited resource, for each trigonometric function requires foresight, and they differ for degrees and radians. Numbers are ultimately represented to humans in decimal, having both an integer and a fractional portion. What is the needed resolution? How many table entries are required to provide that granularity? How many binary bits does it take to reach those boundaries?

Note: *Memory sizing in this section pertains primarily to core functions. While inverse functions are very similar, their differences are covered in section 3.2.*

Example, the sin(θ) function returns a number between -1.0 to +1.0. Therefore, only two bits are required to represent the integer portion, a sign bit and a single integer bit. The fractional portion is another question all together.

The Log2(10) rule, which is ~3.322, shall be used in these and other cases where boundaries are known/qualified. If we want the equivalent of 6 decimal places in fractional binary, 6 * 3.322 = 19.932 or 20 fractional bits. A sign bit, an integer bit, and 20 fractional bits, totaling 22-bits, is all we need for each table entry. The signed Qm.n format notation would be Q1.20.

Note: *An extra bit may be required in the fractional portion to ensure table entries are not duplicated due to the effects of rounding and truncation during instantiation, Q1.21.*

Note: *Q1.23 would be on par with single precision floating point, without the exponent component. However, the designer may be limited by the size of block RAM available in the targeted device.*

Number of entries in the table are and how the address into the table is created are the next questions. This is where the solution for degrees and radians diverge.

Indexing Memory For Degrees (Core Functions)

While we can design for the entire 360°, we only need to account for one quadrant, 0 to 90° in a function, because the other three quadrants are the similar, differing by sign and direction. With respect to degrees, the integer portion requires 7 bits to represent 0 to 89. But how many subentries between each integer portion do we want?

Note: *angles from 90 to 359° are handled by the state machine.*

The best way is to do this is on powers-of-two boundaries: 2, 4, 8, 16, 32 subentries and so on. This will require 1, 2, 3, 4, 5 fractional bits, respectively.

Fractional bits	Fractional numbers
$1=(2^{y-1})$	0.0,0.5
$2=(2^{y-2})$	0.0,0.25,0.5,0.75
$3=(2^{y-3})$	0.0,0.125,0.25,0.375,0.5,0.625,0.75,0.875
$4=(2^{y-4})$	0.0,0.0625,0.125,0.1875,0.25,0.3125,0.375....0.9375
$5=(2^{y-5})$	0.0,0.03125,0.0625,0.09375,0.125,0.15625....0.96875

The resulting memory array is comprised of (90°)*32 subentries = 2880 of Q1.20 data, or 2880 x 22-bits, which is not unreasonable.

The resulting range of degrees in is 0 to 89.96875°, in 2880 0.03125° increments.

Indexing Memory For Radians (Core Functions)

Because a radian is related to π, it's not only a real number but an irrational one as well. Indexing memory requires a binary number as an integer taken from the Qm.n format chosen. Fortunately, the relationship lends itself to a simple solution. Convert the radians number to a fixed-point from a real while rounding away from zero, with enough bits to fit within a power-of-two range, as shown below.

Note: *The binary integer range for the corresponding Qm.n format is not equal, because the fractional bits of the Qm.n bits carry different weight.*

Angle	Relationship	Radians	Binary Range	Bits	Size
0-90°	= $\pi/2$	1.5708	0-12868	14	16K
0-180°	= π	3.1416	0-25736	15	32K
0-270°	= $3\pi/2$	4.7124	0-38604	16	64K
0-360°	= 2π	6.2832	0-51472	16	64K

Like the former example that indexes with degrees, we only need 0 to 90°, or $\pi/2$ radians, in a function, because the other three quadrants are similar, differing by sign and direction. The table input can span 0-12,867; whereas an index value of 12868, which is a quadrant boundary, is handled by the state machine.

Note: *Clarification on this last point: $\pi/2$ rounded toward zero is divided by 12868, indexed from 0 to 12,867. The quadrant boundary of $\pi/2$ (1.5708) rounded away from zero corresponds to an input of 12,868. This approach maintains symmetry from quadrant to quadrant.*

Comparatively, the radians solution has 12,868 entries per quadrant; whereas the degrees solution has 2,880. Consequently the equivalent of 8 decimal places for radians is needed instead of 6 for degrees, 8 * 3.322 = 26.576 or 27 fractional bits.

The resulting memory array is comprised of 12,868 entries of Q1.27 data. 16K by 1-bit is a common block RAM component, but you would need 29 of them (larger FPGA devices can host 2-bits per address resulting in using less block RAM).

Note: *External RAM devices may be a better choice for some radians index solutions provided that the extra latency is not a problem.*

Caveat

Some functions can return numbers approaching infinity, like tangent. When approaching 90 or 270°, the result rapidly increases in size, and becomes undefined when at 90 or 270°.

In cases like this, an extra flag bit within each memory entry is warranted. This bit is set true during instantiation when the result exceeds the magnitude of the operand chosen, or Qm.n range.

Other considerations

- Conversion functions from real to fixed-point and fixed-point to real used in the design and can be found in the in the addendum, "*ConversionPackageV2.vhd*".

  Warning: *the word size, Qm.n, dictates the accuracy of each table entry. While the MATH_REAL.VHD functions return real data values in double precision, when converting from real to fixed-point with the conversion package provided, proper rounding must be selected. Guidance for rounding selection can be found in* "STATE MACHINES IN VHDL *Multipliers* Vol. 2".

- While fast adders can improve performance by increasing the clock speed, only ripple carry adders relying on embedded carry chains are used in this chapter.

- The type of block RAM used is also important, due to speed and latency. The designs in this chapter will use inferred synchronous ROMs. The address is input on the clock edge and data is available for capture on the next clock edge.

Note: *While conceptually simple, a number of things have to happen for these functions to operate, particularly within a VHDL construct.*

3.1 Core Functions

Core functions include sine, cosine, and tangent, along with their reciprocals cosecant, secant, and cotangent. An implementation using a state machine is provided for each function in both degrees and radians. Because these six functions share many common parameters, a single package file was created to support them, named *CoreTablePackage.vhd.*

Note: *a different package file is used for inverse functions.*

3.1.1 Core Table Package File

In VHDL a typical package file contains a header or declarations of the package involving data types, constants, and functions within the body portion. The body portion contains the code for the actual functions. All are visible to designs referencing those constants and functions.

Design Parameters

Degree Based Designs (Core Table Functions):

- Supports 90 entries, for 0 to 89°, and 32 subentries in 0.03125 increments, resulting in 0 to 89.96875 degrees, or 2880 table entries (90*32).
- The actual table only represents one quadrant worth of data. The index into the table for other quadrants is computed by the state machine.
- Input to the state machine spans an entire circle or 0 to 359.96875 degrees, and is input as unsigned Q9.5, or UQ9.5.
- Output of the table depends on the function. Sine and cosine have a range of -1.0 to +1.0 as signed Q1.21. Tangent, cosecant, secant, and cotangent can exceed +/- 1.0 and use an extended format of signed Q6.21. This limits the magnitude of the output to +/-63.999999 (short of ∞).
- Extended entries have an extra flag bit in the table to indicate that the output is at its maximum value or is saturated.
- Standard degree-based constants are prefaced by DEG_, extended are prefaced by XDEG_. Data arrays likewise are deg_da and xdeg_da.

- Quadrant boundaries are represented in degrees as 90, 180, 270, and 360°.

Radian Based Designs (Core Table Functions):

- Support 12,868 entries, representing 1.5708 radians or $\pi/2$.
- The actual table only represents one quadrant worth of data. The index into the table for other quadrants is computed by the state machine.
- Input to the state machine spans an entire circle or 0 to 6.2832 radians or 2π, and is input as unsigned UQ3.13.
- Output of table depends on function. Sine and cosine have a range of -1.0 to +1.0 as signed Q1.27. Tangent, cosecant, secant, and cotangent can exceed +/- 1.0 and use an extended format of signed Q6.27. This limits the magnitude of the output to +/-63.99999999 (again, short of ∞). The expanded fractional portion of from 21 to 27 bits is because 12,868 table entries require more resolution than the 2,880 entries used in degree implementations.
- Extended entries have an extra flag bit in the table to indicate that the output is at its maximum value, or is saturated.
- Standard radians-based constants are prefaced by RAD_, extended are prefaced by XRAD_. Data arrays likewise are rad_da and xrad_da.
- Quadrant boundaries are represented in radians as $\pi/2$, π, $3\pi/2$, and 2π.

Organization Within The Header

- The first group declares constants and data array types used in degree implementations.
- The second group declares constants and data array types used in radians implementations.
- Lastly, support functions, enumeration types used by table initialization functions and declarations of the table initialization functions.

Table Initialization Functions (Core Table Functions)

There are four declarations for table initialization functions, two for degrees and two for radians. Each pair's contents are very similar except one returns an initialized standard data array, the other returns an initialized extended data array.

The purpose of these functions is to create a table of a specific size, index value and format, with an associated output value and format, for a specific trigonometric function over a specific range. These functions are called to initialize an array during instantiation and are typically inferred as synchronous block RAM or ROM.

While these functions can support an entire circle, $360°$ or 2π radians, implementations in this chapter are limited to initializing a table of only one quadrant, $90°$ or $\pi/2$. This is because the substantial memory requirements of a full circle is not justified nor practical.

Following is additional information that has not been covered or would not be clear to the average reader when reviewing source code.

Degree Based Functions (Core Table Functions)

function init_tbl_deg(entries,subentries: natural; func: f_def) return deg_da;

- The parameters *entries* and *subentries* correspond to the integer portion and the fractional portion of the index into block RAM, respectively. In other words, the RAM address. Parameter *func* indicates the type of function, either sine or cosine, which returns a standard data array (no flag bit).
- Data at each address is initialized with the function of the corresponding address, which is the degree parameter.
- The number of fractional bits needed for the number of subentries has to be determined, *frac_width*. This allows the fractional portion to be represented in Q0.n fixed-point format, which is subsequently converted to real and combined with the integer portion. The +1 included in the *frac_width* assignment accounts for the sign bit needed in the Qm.n format of the fractional portion.
- The outer loop corresponds to creating the integer portion of the degrees index and the inner loop corresponds to the fractional portion between each degree.
- Both the integer and fractional are converted to real, combined, then converted to radians, which is then input to the MATH_REAL.VHD function.
- The result of the function is converted back to the required Qm.n format defined in the package declarations, and stored at the current data array location.

function init_tbl_deg(entries,subentries: natural; func: f_def) return xdeg_da;

- The parameters *entries* and *subentries* correspond to the integer portion and the fractional portion of the index into block RAM, respectively. In other words, the RAM address. Parameter *func* indicates the type of function: tangent, cosecant, secant, or cotangent, which returns an extended data array (flag bit).
- Data at each address is initialized with the function of the corresponding address, which is the degree parameter.
- The number of fractional bits needed for the number of subentries has to be determined, *frac_width*. This allows the fractional portion to be represented in Q0.n fixed-point format, which is subsequently converted to real and combined with the integer portion. The +1 included in the

frac_width assignment accounts for the sign bit needed in the Qm.n format of the fractional portion.

- The outer loop corresponds to creating the integer portion of the degrees index, and the inner loop corresponds to the fractional portion between each degree.
- Both the integer and fractional are converted to real, combined, then converted to radians, which is then input to the MATH_REAL.VHD function.
- The result of the function is compared to the maximum or saturated value of the output Qm.n format. If greater, the value stored at the data array location is set to its maximum and the associated flag bit is set. If not, the result is converted back to the required Qm.n format defined in the package declarations and stored at that location in the array.

Radians Based Functions (Core Table Functions)

function init_tbl_rad(pi_radians: real; entries: natural; func: f_def) return rad_da;

- Parameter *pi_radians* corresponds to range of pi: $\pi/2$, π, $3\pi/2$, or 2π, or quadrant boundaries. Parameter *entries* represents the number of increments are over that range and is typically an offset representation of radians, 12,868 (0-12,867) for $\pi/2$ radians (**rounded toward zero**), and so on, and is an integer value used as the RAM address to the data array. In other words, parameters of $\pi/2$ and 12868, will result in dividing $\pi/2$ by 12,868 and each memory location will be set to $((\pi/2)/12,868)*$RAM address. Parameter *func* indicates the type of function, either sine or cosine, which returns a standard data array (no flag bit).
- Data at each address is initialized with the function of the corresponding address, which is the radians parameter.
- The result of the function is converted back to the required Qm.n format defined in the package declarations and stored at the current array location.

function init_tbl_rad(pi_radians: real; entries: natural; func: f_def) return xrad_da;

- Parameter *pi_radians* corresponds to range of pi: $\pi/2$, π, $3\pi/2$, or 2π, or quadrant boundaries. Parameter *entries* represents the number of increments are over that range and is typically an offset representation of radians, 12,868 (0-12,867) for $\pi/2$ radians (**rounded toward zero**), and so on, and is an integer value used as the RAM address to the data array. In other words, parameters of $\pi/2$ and 12868, will result in dividing $\pi/2$ by 12,868 and each memory location will be set to $((\pi/2)/12,868)*$RAM address. Parameter *func* indicates the type of function: tangent, cosecant, secant, or cotangent, which returns an extended data array (flag bit).
- Data at each address is initialized with the function of the corresponding address, which is the radians parameter.
- The result of the function is compared to the maximum or saturated value of the output Qm.n format. If greater, the value stored at the array location is set to its maximum and the associated flag bit is set. If not, the result is converted back to the required Qm.n format defined in the package declarations and is stored at that location in the array.

```
------------------------------------------------------------
--  CoreTablePackage.vhd
------------------------------------------------------------
library IEEE;
use IEEE.std_logic_1164.all;
use IEEE.numeric_std.all;
use IEEE.math_real.all;
use work.ConversionPackageV2.all;


--  package header
package CoreTablePackage is


-- Constants for degrees input implementation. ENTRIES and SUBENTRIES
combined
-- correspond to the number of table entries for 90 degrees, integer portion and
-- fractional portion, respectively. Data input is scaled to span the entire
-- 360 degrees, 9-bit for integer and 5-bits for fractional, unsigned fixed-point.
constant DEG_ENTRIES: natural := 90; -- 90 degrees
constant DEG_SUBENTRIES: natural := 32; -- 32 point between each degree
constant DEG_DI_INT_SIZ: natural := 9; -- supports 360 degrees
constant DEG_DI_FRAC_SIZ: natural := 5; -- bits for 32 subentries
constant DEG_DI_SIZ: natural := DEG_DI_INT_SIZ + DEG_DI_FRAC_SIZ;
-- constants for quadrant boundaries
constant \0 DEGS\: unsigned(DEG_DI_SIZ-1 downto 0)   :=
    B"000000000_00000";
constant \90 DEGS\: unsigned(DEG_DI_SIZ-1 downto 0)  :=
    B"001011010_00000";
constant \180 DEGS\: unsigned(DEG_DI_SIZ-1 downto 0) :=
    "010110100_00000";
constant \270 DEGS\: unsigned(DEG_DI_SIZ-1 downto 0) :=
    B"100001110_00000";
constant \360 DEGS\: unsigned(DEG_DI_SIZ-1 downto 0) :=
    B"101101000_00000";
-- signed fixed-point formatting constants for standard output data
constant DEG_DO_INT_SIZ: natural := 1; -- integer size
constant DEG_DO_FRAC_SIZ: natural := 21; -- fractional size
constant DEG_DO_SIZ: natural := 1+DEG_DO_INT_SIZ+DEG_DO_FRAC_SIZ;
-- constants for output     data
constant DEG_PLUS1: signed(DEG_DO_SIZ-1 downto 0) :=
    RealToQmn(1.0,DEG_DO_INT_SIZ,DEG_DO_FRAC_SIZ,TZERO);
constant DEG_MINUS1: signed(DEG_DO_SIZ-1 downto 0) :=
    RealToQmn(-1.0,DEG_DO_INT_SIZ,DEG_DO_FRAC_SIZ,TZERO);
-- standard data array
type deg_da is array (natural range <>) of signed(DEG_DO_SIZ-1 downto 0);
```

```
-- signed fixed-point formatting constants for extended output
constant XDEG_DO_INT_SIZ: natural := 6; -- integer size
constant XDEG_DO_FRAC_SIZ: natural := 21; -- fractional size
constant XDEG_DO_SIZ: natural :=
    1+XDEG_DO_INT_SIZ+XDEG_DO_FRAC_SIZ;
constant XDEG_TBL_WIDTH: natural := 1+XDEG_DO_SIZ; -- extra flag bit
-- constants for output data
constant XDEG_PLUS1: signed(XDEG_DO_SIZ-1 downto 0) :=
    RealToQmn(1.0,XDEG_DO_INT_SIZ,XDEG_DO_FRAC_SIZ,TZERO);
constant XDEG_MINUS1: signed(XDEG_DO_SIZ-1 downto 0) :=
    RealToQmn(-1.0,XDEG_DO_INT_SIZ,XDEG_DO_FRAC_SIZ,TZERO);
-- extended data array
type xdeg_da is array (natural range <>) of  signed(XDEG_TBL_WIDTH-1
    downto 0);
-- Constants for radians input implementation. RAD_ENTRIES corresponds
-- to the number of table entries for pi/2 radians. Data input is scaled to be
-- 2pi radians, full circle, unsigned fixed-point.
constant RAD_ENTRIES: natural := 12868; -- binary equivalent to Qm.n 1.5708
    radians equal to ~pi/2
constant RAD_DI_INT_SIZ: natural := 3; -- integer size
constant RAD_DI_FRAC_SIZ: natural := 13; -- fractional size
constant RAD_DI_SIZ: natural := RAD_DI_INT_SIZ+RAD_DI_FRAC_SIZ;
-- constants for quadrant boundaries
constant \0\: unsigned(RAD_DI_SIZ-1 downto 0)  := (others=>'0');
constant \pi/2\: unsigned(RAD_DI_SIZ-1 downto 0) :=
    TO_UNSIGNED(12868,RAD_DI_SIZ);
constant \pi\: unsigned(RAD_DI_SIZ-1 downto 0) :=
    TO_UNSIGNED(25736,RAD_DI_SIZ);
constant \3pi/2\: unsigned(RAD_DI_SIZ-1 downto 0) :=
    TO_UNSIGNED(38604,RAD_DI_SIZ);
constant \2pi\: unsigned(RAD_DI_SIZ-1 downto 0) :=
    TO_UNSIGNED(51472,RAD_DI_SIZ);
-- signed fixed-point formatting constants for standard output data
constant RAD_DO_INT_SIZ: natural := 1; -- integer size
constant RAD_DO_FRAC_SIZ: natural := 27; -- fractional size
constant RAD_DO_SIZ: natural := 1+RAD_DO_INT_SIZ+RAD_DO_FRAC_SIZ;
-- constants for output data
constant RAD_PLUS1: signed(RAD_DO_SIZ-1 downto 0) :=
    RealToQmn(1.0,RAD_DO_INT_SIZ,RAD_DO_FRAC_SIZ,TZERO);
constant RAD_MINUS1: signed(RAD_DO_SIZ-1 downto 0) :=
    RealToQmn(-1.0,RAD_DO_INT_SIZ,RAD_DO_FRAC_SIZ,TZERO);
-- standard data array
type rad_da is array (natural range <>) of signed(RAD_DO_SIZ-1 downto 0);
-- signed fixed-point formatting constants for extended output data
```

```vhdl
constant XRAD_DO_INT_SIZ: natural := 6; -- integer size
constant XRAD_DO_FRAC_SIZ: natural := 27; -- fractional size
constant XRAD_DO_SIZ: natural :=
    1+XRAD_DO_INT_SIZ+XRAD_DO_FRAC_SIZ;
constant XRAD_TBL_WIDTH: natural := 1 + XRAD_DO_SIZ; -- extra flag bit
-- constants for output data
constant XRAD_PLUS1: signed(XRAD_DO_SIZ-1 downto 0) :=
    RealToQmn(1.0,XRAD_DO_INT_SIZ,XRAD_DO_FRAC_SIZ,TZERO);
constant XRAD_MINUS1: signed(XRAD_DO_SIZ-1 downto 0) :=
    RealToQmn(-1.0,XRAD_DO_INT_SIZ,XRAD_DO_FRAC_SIZ,TZERO);
-- extended data array
type xrad_da is array (natural range <>) of signed(XRAD_TBL_WIDTH-1
    downto 0);


-- support functions
-- returns the maximum positive qmn number for word size
function max_pos_qmn(n: integer) return signed;


-- table initialization function types
type f_def is (SINF,COSF,TANF,CSCF,SECF,COTF);
-- core trig functions with reciprocals, six in all
-- initialize table functions for degree based table
function init_tbl_deg(entries,subentries: natural; func: f_def) return deg_da;
function init_tbl_deg(entries,subentries: natural; func: f_def) return xdeg_da;
-- initialize table functions for radians based table
function init_tbl_rad(pi_radians: real; entries: natural; func: f_def) return rad_da;
function init_tbl_rad(pi_radians: real; entries: natural; func: f_def) return xrad_da;

end;


--    package body
package body CoreTablePackage is

-- support functions
-- returns the maximum positive qmn number for word size
function max_pos_qmn(n: integer) return signed is
variable qmn: signed(n-1 downto 0):= (others=>'1');
begin
    qmn(qmn'high) :='0'; -- clear sign bit
    return(qmn);
end function;
--
--    Initialization functions (based on degrees)
--
```

-- Initialize table of any size from 0 to 360 degrees
--
-- Entries represent the integer granularity in degrees. Subentries represent
-- the fractional granularity. The combination of entries * subentries represents
-- the number of table entries. This function only supports Sine and Cosine, the
-- standard degrees data array is returned, no flag bit.

```
function init_tbl_deg(entries,subentries: natural; func: f_def) return deg_da is
variable int_real,frac_real: real := 0.0;
constant frac_width: natural := natural(ceil(log2(real(subentries))))+1;
variable frac_signed: signed(frac_width-1 downto 0):= (others=>'0');
variable degrees,radians,result: real := 0.0;
variable da: deg_da(0 to (entries*subentries)-1) := (others=>(others=>'0'));
begin
    -- entries correspond to integer granularity
    for i in 0 to entries-1 loop
        int_real := real(i);
        -- sub entries correspond to fractional granularity
        for j in 0 to subentries-1 loop
            -- convert subentries into positive signed number
            frac_signed := TO_SIGNED(j,integer(frac_width));
            -- convert subentries to fractional real number
            frac_real := QmnToReal((frac_signed),0,integer(frac_width-1));
            -- combine integer and fractional portions
            degrees := int_real + frac_real;
            -- convert from degrees to radians
            radians := (degrees * MATH_PI)/180.0;
            -- employ math_real trig functions
            case func is
                when SINF => result := SIN(radians);
                when COSF => result := COS(radians);
                when others => report "Function not supported!" severity failure;
            end case;
            -- convert to Qm.n data format and place in table entry
            da((i*subentries)+j) :=
            RealToQmn(result,DEG_DO_INT_SIZ,DEG_DO_FRAC_SIZ,TZERO);
        end loop;
    end loop;

    return(da);
end function;
```

-- This function supports Tangent, Cosecant, Secant, and Cotangent. The
-- returned data array contains a flag bit, denoted by the leading x. When the flag
-- bit is set the data at location is set to its maximum value.

```
function init_tbl_deg(entries,subentries: natural; func: f_def) return xdeg_da is
```

```vhdl
variable int_real,frac_real: real := 0.0;
constant frac_width: natural := natural(ceil(log2(real(subentries))))+1;
variable frac_signed: signed(frac_width-1 downto 0):= (others=>'0');
variable degrees,radians,results: real := 0.0;
variable xda: xdeg_da(0 to (entries*subentries)-1) := (others=>(others=>'0'));
begin
    -- entries correspond to integer granularity
    for i in 0 to entries-1 loop
        int_real := real(i);
        -- sub entries correspond to fractional granularity
        for j in 0 to subentries-1 loop
            -- convert subentries into positive signed number
            frac_signed := TO_SIGNED(j,integer(frac_width));
            -- convert subentries to fractional real number
            frac_real := QmnToReal((frac_signed),0,integer(frac_width-1));
            -- combine integer and fractional portions
            degrees := int_real + frac_real;
            -- convert from degrees to radians
            radians :=  (degrees * MATH_PI)/180.0;
            -- employ math_real trig function
            case func is
                when TANF => results := TAN(radians);
                when CSCF => if(SIN(radians) = 0.0) then
                                    results := 0.0;
                             else
                                    results := 1.0/SIN(radians);
                             end if;
                when SECF => if(COS(radians) = 0.0) then
                                    results := 0.0;
                             else
                                    results := 1.0/COS(radians);
                             end if;
                when COTF => if(TAN(radians) = 0.0) then
                                    results := 0.0;
                             else
                                    results := 1.0/TAN(radians);
                             end if;
                when others => report "Function not supported!" severity failure;
            end case;
            -- check boundary for saturation
            if(results >= QmnToReal(max_pos_qmn(XDEG_DO_SIZ),
                XDEG_DO_INT_SIZ,XDEG_DO_FRAC_SIZ)) then
                -- set flag bit and assign the highest positive number
                xda((i*subentries)+j) := '1'&max_pos_qmn(XDEG_DO_SIZ);
```

```
                else
                    -- convert to Qm.n data format and place in table entry along with
                    -- cleared flag bit
                    xda((i*subentries)+j) := '0'&RealToQmn(results,
                        XDEG_DO_INT_SIZ,XDEG_DO_FRAC_SIZ,TZERO);
                end if;
            end loop;
        end loop;

        return(xda);
        end function;
--
--  Initialization functions (based on radians)
--
--  Initialize sine table of any size from 0 to 2pi radians
--
--  pi_radians is on boundaries of pi/2, pi, 3pi/2, or 2pi. Entries represent
--  the number of table entries for the value of pi_radians. This function only
--  supports Sine and Cosine, the   standard radians data array is returned, no flag
--  bit.
function init_tbl_rad(pi_radians: real; entries: natural; func: f_def) return rad_da is
variable radians,results: real := 0.0;
variable da: rad_da(0 to entries-1) := (others=>(others=>'0'));
begin
        -- entries correspond table entries
        for i in 0 to entries-1 loop
            -- compute fractional radians based on current entry
            radians := (pi_radians/real(entries)) * real(i);
            -- employ math_real sine function
            case func is
                when SINF => results := SIN(radians);
                when COSF => results := COS(radians);
                when others => report "Function not supported!" severity failure;
            end case;
            -- convert to Qm.n data format and place in table entry
            da(I) := RealToQmn(results,
                RAD_DO_INT_SIZ,RAD_DO_FRAC_SIZ,TZERO);
        end loop;

        return(da);
end function;
--  This function supports Tangent, Cosecant, Secant, and Cotangent. The
--  returned data array contains a flag bit, denoted by the leading x. When the flag
--  bit is set the data at location is set to its maximum value.
```

```vhdl
function init_tbl_rad(pi_radians: real; entries: natural; func: f_def) return xrad_da is
variable radians,results: real := 0.0;
variable xda: xrad_da(0 to entries-1) := (others=>(others=>'0'));
begin
    -- entries correspond table entries
    for i in 0 to entries-1 loop
        -- compute fractional radians based on current entry
        radians := (pi_radians/real(entries)) * real(i);
        -- employ math_real sine function
        case func is
            when TANF => results := TAN(radians);
            when CSCF => if(SIN(radians) = 0.0) then
                            results := 0.0;
                        else
                            results := 1.0/SIN(radians);
                        end if;
            when SECF => if(COS(radians) = 0.0) then
                            results := 0.0;
                        else
                            results := 1.0/COS(radians);
                        end if;
            when COTF => if(TAN(radians) = 0.0) then
                            results := 0.0;
                        else
                            results := 1.0/TAN(radians);
                        end if;
            when others => report "Function not supported!" severity failure;
        end case;
        -- check boundary for saturation
        if(results >= QmnToReal(max_pos_qmn(XRAD_DO_SIZ),
            XRAD_DO_INT_SIZ,XRAD_DO_FRAC_SIZ)) then
            -- set flag bit and assign the highest positive number
            xda(i) := '1'&max_pos_qmn(XRAD_DO_SIZ);
        else
            -- convert to Qm.n data format and place in table entry along with
            -- cleared flag bit
            xda(i) := '0'&RealToQmn(results,
                XRAD_DO_INT_SIZ,XRAD_DO_FRAC_SIZ,TZERO);
        end if;
    end loop;

    return(xda);
end function;
end;
```

3.1.2 Core State Machines

Trigonometric functions are periodic and symmetrical from quadrant to quadrant, except that direction and polarity change between quadrants. Because of this, values from only one quadrant are needed to create all four quadrants, hence only a look-up table containing one quadrant of data for a given function, aptly illustrated below in Figure 2.

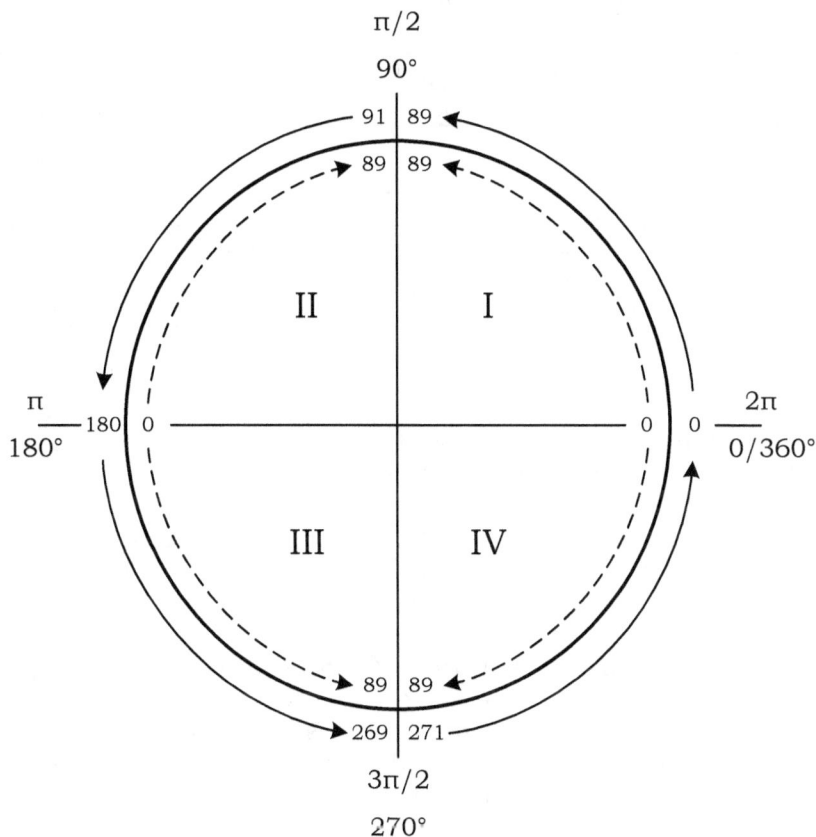

Figure 2

For purpose of simplicity, in this explanation degrees are used only in integer form. A data table (RAM block) with 90 entries is initialized with an address granularity of 1°, corresponding to address values 0 to 89. If the table is indexed as delineated by the dotted lines (circle interior) relative to the actual position (circle exterior), the magnitude of the function is the same for each quadrant. The state machine further evaluates the quadrant to determine the sign of the magnitude.

Quadrant boundaries 90° and 270° do not have a data table entries. Instead, data is asserted by the state machine with constants -- values differing from function to function. There are also cases where the data value is undefined at a given quadrant boundary, as shown in the table below. A status signal is driven by the state machine to indicate this condition.

Functions	Quadrant Boundaries -- versus -- Data Sources			
	0/360°/ 2π	90°/ π/2	180°/ π	270°/ 3π/2
Sine	Table	+1.0	Table	-1.0
Cosine	Table	0	Table	0
Tangent	Table	Undefined	Table	Undefined
Cosecant	Undefined	+1.0	Undefined	-1.0
Secant	Table	Undefined	Table	Undefined
Cotangent	Undefined	0	Undefined	0

Using this paradigm, the structure and number of states is identical for all core functions and their reciprocals. Differences include what the table is initialized with, what values are asserted at 90 and 270°, and those boundaries that are undefined.

Two designs are provided for each function, one implemented with degrees as an input and one with radians as an input. Figure 3 illustrates the basic state diagrams. Since all designs are essentially identical, one fully coded design is provided for each type. Similar types only show the differences in code. Sine using degrees and radians embodies all source code, while the cosine code only provides the differences. Likewise, tangent using degrees and radians contains all source code, while cosecant, secant, and cotangent only shown the differences.

The reader could easily combine several designs into a single state machine provided adequate memory resources are available in the targeted device. Sine and cosine by managing the table index only.

Performance information is provided. The Xilinx Spartan 3E was chosen for all degree input implementation due to their economic advantage, a Virtex 5 was chosen for the radians input implementation due to its memory resources.

Note: *2 to 5 clock cycles are required to execute each function.*

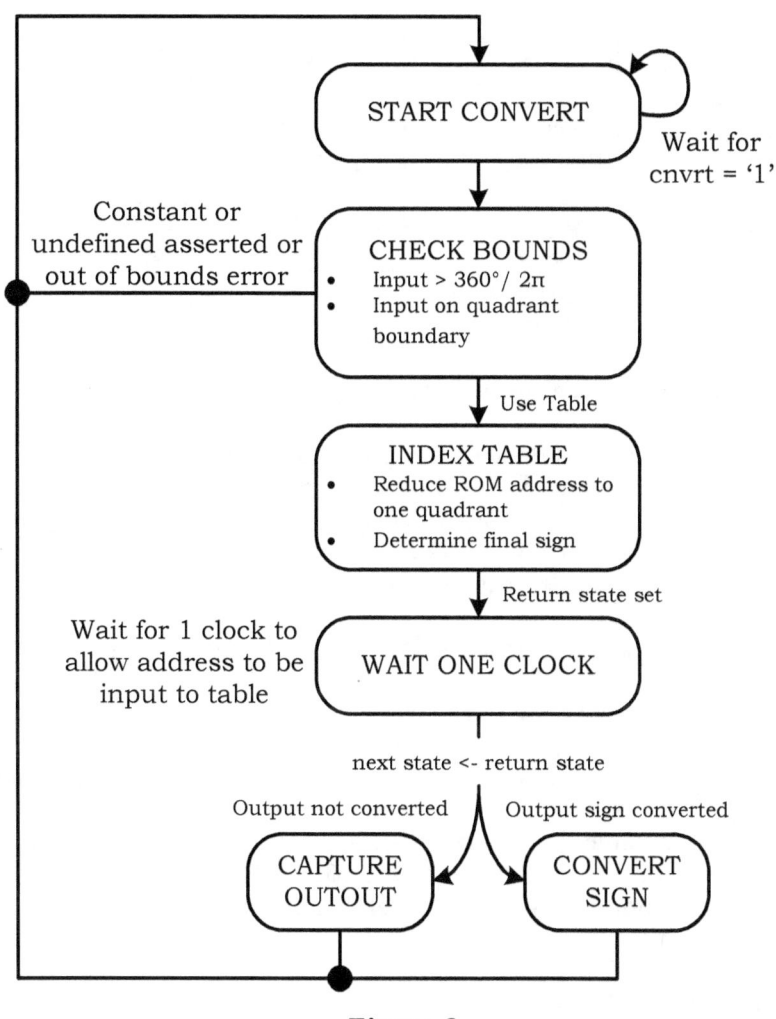

Figure 3

Function	Implementation	Targeted Part	Clock Speed
Sine	Degrees	XC3S500E	148MHz
	Radians	XC5VLX30	196MHz
Cosine	Degrees	XC3S500E	147MHz
	Radians	XC5VLX30	202MHz
Tangent	Degrees	XC3S500E	142MHz
	Radians	XC5VLX30	198MHz
Cosecant	Degrees	XC3S500E	139MHz
	Radians	XC5VLX30	171MHz
Secant	Degrees	XC3S500E	142MHz
	Radians	XC5VLX30	198MHz
Cotangent	Degrees	XC3S500E	137MHz
	Radians	XC5VLX30	184MHz

Recap on state machine interface signals

- All degree based designs support UQ9.5 for unsigned 360.0° input. Sine and cosine support signed Q1.21 output for -1.0 to +1.0; whereas tangent, cosecant, secant, and cotangent support Q6.21 for -63.999999 to +63.999999, and also have two additional status bits *ovr* (output saturated) and *udf* (undefined).
- All radians based designs support U3.13 for unsigned 6.2832 radians input. Sine and cosine support signed Q1.27 output for -1.0 to +1.0; whereas tangent, cosecant, secant, and cotangent support Q6.27 for -63.99999999 to +63.99999999, and also have two additional status bits *ovr* (output saturated) and *udf* (undefined).
- Both implementations have an *err* status bit indicating the input has exceeded the 360° or 6.2832 radians limit.
- Synchronous ROM (using block RAM) is inferred in every design.

Possible Improvements

- The use of faster adders. In all cases, the adder that does a two's complement of the ROM output is the limiting factor of the design.

3.1.2.1   Sine Functions (source code)

```
-------------------------------------------------------------------
--   SineUsingDegrees.vhd (degrees as input)
-------------------------------------------------------------------
library IEEE;
use IEEE.std_logic_1164.all;
use IEEE.numeric_std.all;
use IEEE.math_real.all;
-- user packages
use work.ConversionPackageV2.all;
use work.CoreTablePackage.all;

entity SineUsingDegrees is
--   Input and output fixed-point formats are defined in CoreTablePackage.vhd
port
(
    clk,rst: in std_logic; -- system clock and reset (synchronous)
    -- inputs
    cnvrt: in std_logic; -- initiate sine function
    theta: in unsigned(DEG_DI_SIZ-1 downto 0); -- theta in degrees
    -- outputs
    rdy: out std_logic; -- data ready
    err: out std_logic; -- error, input out of range
    func_of_angle: out signed(DEG_DO_SIZ-1 downto 0)
);
end SineUsingDegrees;

architecture RTL of SineUsingDegrees is

--   declared constants
constant TABLE_SIZ: natural := DEG_ENTRIES*DEG_SUBENTRIES;
constant ROM_ARRAY: deg_da(0 to TABLE_SIZ-1) :=
    init_tbl_deg(DEG_ENTRIES,DEG_SUBENTRIES,SINF);
constant ROM_ADR_SIZ: natural := natural(ceil(log2(real(TABLE_SIZ))));
--   declared signals
signal degrees: unsigned(DEG_DI_SIZ-1 downto 0) := (others=>'0');
signal rom_adr: unsigned(DEG_DI_SIZ-1 downto 0) := (others=>'0');
signal norm_rom_adr: unsigned(ROM_ADR_SIZ-1 downto 0) := (others=>'0');
signal rom_out: signed(DEG_DO_SIZ-1 downto 0) := (others=>'0');
--   state machine enumeration list
type sm_def is
(
    RESET,START_CNVRT,CHECK_BOUNDS,INDEX_TABLE,
```

```vhdl
        WAIT_ONE_CLK,CAPTURE_OUTPUT,CONVERT_SIGN
);
signal state, ret_state: sm_def := RESET;

begin
---------------------------------------------
--   Sine Function state machine (using degrees)
---------------------------------------------
process(rst,clk)
begin
    if(rst='1') then
        -- outputs
        rdy <= '0';err <= '0';
        func_of_angle <= (others=>'0');
        -- local
        rom_adr <= (others=>'0');
        degrees <= (others=>'0');
        -- states
        state <= RESET;
        ret_state <= RESET;
    elsif rising_edge(clk) then
        --
        --   state machine body
        --
        case state is
            -- reset state
            when RESET =>
                state <= START_CNVRT;
            -- wait for signal to start conversion
            when START_CNVRT =>
                if(cnvrt = '1') then
                    rdy <= '0';err <= '0';
                    degrees <= theta;
                    state <= CHECK_BOUNDS;
                end if;
            -- check basic bounds and directly assign boundaries of quadrants.
            -- theta of 0, 180, and 360 are handled indirectly.
            when CHECK_BOUNDS =>
                if(degrees > \360 DEGS\) then
                    rdy <= '1';err <= '1';
                    state <= START_CNVRT;
                elsif(degrees = \270 DEGS\) then
                    func_of_angle <= DEG_MINUS1;
                    rdy <= '1';
```

```vhdl
                    state <= START_CNVRT;
              elsif(degrees = \90 DEGS\) then
                    func_of_angle <= DEG_PLUS1;
                    rdy <= '1';
                    state <= START_CNVRT;
                else
                    state <= INDEX_TABLE;
                end if;
            -- create index into ROM
            when INDEX_TABLE =>
                if(degrees > \270 DEGS\) then
                    rom_adr <= \90 DEGS\ - (degrees - \270 DEGS\);
                    ret_state <= CONVERT_SIGN;
                elsif(degrees > \180 DEGS\) then
                    rom_adr <= degrees - \180 DEGS\;
                    ret_state <= CONVERT_SIGN;
                elsif(degrees > \90 DEGS\) then
                    rom_adr <= \90 DEGS\ - (degrees - \90 DEGS\);
                    ret_state <= CAPTURE_OUTPUT;
                else
                    rom_adr <= degrees;
                    ret_state <= CAPTURE_OUTPUT;
                end if;
                state <= WAIT_ONE_CLK;
            when WAIT_ONE_CLK =>
                state <= ret_state;
            when CAPTURE_OUTPUT =>
                func_of_angle <= rom_out;
                rdy <= '1';
                state <= START_CNVRT;
            when CONVERT_SIGN =>
                func_of_angle <= (not rom_out) + 1;
                rdy <= '1';
                state <= START_CNVRT;
            when others =>
                state <= RESET;
        end case;
    end if;
end process;
-- normalize rom address
norm_rom_adr <= rom_adr(norm_rom_adr'high downto 0);
```

```vhdl
-- synchronous ROM
process(clk)
begin
    if rising_edge(clk) then
        rom_out <= ROM_ARRAY(TO_INTEGER(norm_rom_adr));
    end if;
end process;

end RTL;
```

```
-------------------------------------------------------------------------------
--   SineUsingRadians.vhd (radians as input)
-------------------------------------------------------------------------------
library IEEE;
use IEEE.std_logic_1164.all;
use IEEE.numeric_std.all;
use IEEE.math_real.all;
-- user packages
use work.ConversionPackageV2.all;
use work.CoreTablePackage.all;

entity SineUsingRadians is
--   Input and output fixed-point formats are defined in CoreTablePackage.vhd
port
(
    clk,rst: in std_logic; -- system clock and reset (synchronous)
    -- inputs
    cnvrt: in std_logic; -- initiate sine function
    theta: in unsigned(RAD_DI_SIZ-1 downto 0); -- theta in radians
    -- outputs
    rdy: out std_logic; -- data ready
    err: out std_logic; -- error, input out of range
    func_of_angle: out signed(RAD_DO_SIZ-1 downto 0)
);
end SineUsingRadians;

architecture RTL of SineUsingRadians is

--   declared constants
constant TABLE_SIZ: natural := RAD_ENTRIES;
constant ROM_ARRAY: rad_da(0 to TABLE_SIZ-1) :=
    init_tbl_rad(MATH_PI_OVER_2,TABLE_SIZ,SINF);
constant ROM_ADR_SIZ: natural := natural(ceil(log2(real(TABLE_SIZ))));
--   declared signals
signal angle: unsigned(RAD_DI_SIZ-1 downto 0) := (others=>'0');
signal rom_adr: unsigned(RAD_DI_SIZ-1 downto 0) := (others=>'0');
signal norm_rom_adr: unsigned(ROM_ADR_SIZ-1 downto 0) := (others=>'0');
signal rom_out: signed(RAD_DO_SIZ-1 downto 0) := (others=>'0');
--   state machine enumeration list
type sm_def is
(
    RESET,START_CNVRT,CHECK_BOUNDS,INDEX_TABLE,
    WAIT_ONE_CLK,CAPTURE_OUTPUT,CONVERT_SIGN
);
```

```vhdl
signal state, ret_state: sm_def := RESET;

begin
-----------------------------------------------
--   Sine Function state machine (using Radians)
-----------------------------------------------
process(rst,clk)
begin
    if(rst='1') then
        -- outputs
        rdy <= '0';err <= '0';
        func_of_angle <= (others=>'0');
        -- local
        rom_adr <= (others=>'0');
        -- states
        state <= RESET;
        ret_state <= RESET;
    elsif rising_edge(clk) then
        --   state machine body
        case state is
            -- reset state
            when RESET =>
                state <= START_CNVRT;
            -- wait for signal to start conversion
            when START_CNVRT =>
                if(cnvrt = '1') then
                    rdy <= '0';err <= '0';
                    angle <= theta;
                    state <= CHECK_BOUNDS;
                end if;
            -- check basic bounds and directly assign boundaries of pi.
            -- angles of 0, pi, and 2pi are handled indirectly.
            when CHECK_BOUNDS =>
                if(angle > \2pi\) then
                    rdy <= '1';err <= '1';
                    state <= START_CNVRT;
                elsif(angle = \3pi/2\) then
                    func_of_angle <= RAD_MINUS1;
                    rdy <= '1';
                    state <= START_CNVRT;
                elsif(angle = \pi/2\) then
                    func_of_angle <= RAD_PLUS1;
                    rdy <= '1';
                    state <= START_CNVRT;
```

```vhdl
                else
                    state <= INDEX_TABLE;
                end if;
            -- created index into ROM
            when INDEX_TABLE =>
                if(angle > \3pi/2\) then
                    rom_adr <= \pi/2\ - (angle - \3pi/2\);
                    ret_state <= CONVERT_SIGN;
                elsif(angle > \pi\) then
                    rom_adr <= angle - \pi\;
                    ret_state <= CONVERT_SIGN;
                elsif(angle > \pi/2\) then
                    rom_adr <= \pi/2\ - (angle - \pi/2\);
                    ret_state <= CAPTURE_OUTPUT;
                else
                    rom_adr <= angle;
                    ret_state <= CAPTURE_OUTPUT;
                end if;
                state <= WAIT_ONE_CLK;
            when WAIT_ONE_CLK =>
                state <= ret_state;
            when CAPTURE_OUTPUT =>
                func_of_angle <= rom_out;
                rdy <= '1';
                state <= START_CNVRT;
            when CONVERT_SIGN =>
                func_of_angle <= (not rom_out) + 1;
                rdy <= '1';
                state <= START_CNVRT;
            when others =>
                state <= RESET;
        end case;
    end if;
end process;
-- normalize rom address
norm_rom_adr <= rom_adr(norm_rom_adr'high downto 0);
-- synchronous ROM
process(clk)
begin
    if rising_edge(clk) then
        rom_out <= ROM_ARRAY(TO_INTEGER(norm_rom_adr));
    end if;
end process;
end RTL;
```

3.1.2.2 Cosine Functions (source code)

Note: *Only differences are shown, all other source code is identical to the corresponding sine functions, both degree and radians based.*

```
----------------------------------------------------------------
--  CosineUsingDegrees.vhd (degrees as input)
----------------------------------------------------------------

entity CosineUsingDegrees is

end CosineUsingDegrees;

architecture RTL of CosineUsingDegrees is

--  declared constants

constant ROM_ARRAY: deg_da(0 to TABLE_SIZ-1) :=
    init_tbl_deg(DEG_ENTRIES,DEG_SUBENTRIES,COSF);

begin
-----------------------------------------------
--  Cosine Function state machine (using degrees)
-----------------------------------------------
process(rst,clk)
begin
    if(rst='1') then

    elsif rising_edge(clk) then
        --  state machine body
        case state is

            -- check basic bounds and directly assign boundaries of quadrants.
            -- theta of 0, 180, and 360 are handled indirectly.
            when CHECK_BOUNDS =>
                if(degrees > \360 DEGS\) then
                    rdy <= '1';err <= '1';
                    state <= START_CNVRT;
                elsif(degrees = \270 DEGS\) then
                    func_of_angle <= (others=>'0'); -- zero
                    rdy <= '1';
                    state <= START_CNVRT;
                elsif(degrees = \90 DEGS\) then
```

```vhdl
                    func_of_angle <= (others=>'0'); -- zero
                    rdy <= '1';
                    state <= START_CNVRT;
                else
                    state <= INDEX_TABLE;
                end if;
            -- create index into ROM
            when INDEX_TABLE =>
                if(degrees > \270 DEGS\) then
                    rom_adr <= \90 DEGS\ - (degrees - \270 DEGS\);
                    ret_state <= CAPTURE_OUTPUT;
                elsif(degrees > \180 DEGS\) then
                    rom_adr <= degrees - \180 DEGS\;
                    ret_state <= CONVERT_SIGN;
                elsif(degrees > \90 DEGS\) then
                    rom_adr <= \90 DEGS\ - (degrees - \90 DEGS\);
                    ret_state <= CONVERT_SIGN;
                else
                    rom_adr <= degrees;
                    ret_state <= CAPTURE_OUTPUT;
                end if;
                state <= WAIT_ONE_CLK;
                .

        end case;
    end if;
end process;
.
end RTL;
```

```
---------------------------------------------------------------------------
--  CosineUsingRadians.vhd (radians as input)
---------------------------------------------------------------------------

.

entity CosineUsingRadians is

.

end CosineUsingRadians;

architecture RTL of CosineUsingRadians is

--  declared constants

.

constant ROM_ARRAY: rad_da(0 to TABLE_SIZ-1) :=
    init_tbl_rad(MATH_PI_OVER_2,TABLE_SIZ,COSF);

.

begin
-------------------------------------------------
--  Cosine Function state machine (using Radians)
-------------------------------------------------
process(rst,clk)
begin
    if(rst='1') then

        .

    elsif rising_edge(clk) then
        --  state machine body
        case state is

            .

            -- check basic bounds and directly assign boundaries of pi.
            -- angles of 0, pi, and 2pi are handled indirectly.
            when CHECK_BOUNDS =>
                if(angle > \2pi\) then
                    rdy <= '1';err <= '1';
                    state <= START_CNVRT;
                elsif(angle = \3pi/2\) then
                    func_of_angle <= (others=>'0'); --zero
                    rdy <= '1';
                    state <= START_CNVRT;
                elsif(angle = \pi/2\) then
                    func_of_angle <= (others=>'0'); -- zero;
                    rdy <= '1';
                    state <= START_CNVRT;
                else
                    state <= INDEX_TABLE;
                end if;
```

```
            -- created index into ROM
            when INDEX_TABLE =>
                if(angle > \3pi/2\) then
                    rom_adr <= \pi/2\ - (angle - \3pi/2\);
                    ret_state <= CAPTURE_OUTPUT;
                elsif(angle > \pi\) then
                    rom_adr <= angle - \pi\;
                    ret_state <= CONVERT_SIGN;
                elsif(angle > \pi/2\) then
                    rom_adr <= \pi/2\ - (angle - \pi/2\);
                    ret_state <= CONVERT_SIGN;
                else
                    rom_adr <= angle;
                    ret_state <= CAPTURE_OUTPUT;
                end if;
                state <= WAIT_ONE_CLK;

        end case;
    end if;
end process;

end RTL;
```

3.1.2.3   Tangent Functions (source code)

```
--------------------------------------------------------------------
--  TangentUsingDegrees.vhd   (degrees as input)
--------------------------------------------------------------------
library IEEE;
use IEEE.std_logic_1164.all;
use IEEE.numeric_std.all;
use IEEE.math_real.all;
-- user packages
use work.ConversionPackageV2.all;
use work.CoreTablePackage.all;

entity TangentUsingDegrees is
--   Input and output fixed-point formats are defined in CoreTablePackage.vhd
port
(
    clk,rst: in std_logic; -- system clock and reset (synchronous)
    -- inputs
    cnvrt: in std_logic; -- initiate sine function
    theta: in unsigned(DEG_DI_SIZ-1 downto 0); -- theta in degrees
    -- outputs
    rdy: out std_logic; -- data ready
    err: out std_logic; -- error, input out of range
    ovr: out std_logic; -- output at its maximum range, saturated
    udf: out std_logic; -- undefined at 90 or 270 degree boundaries
    func_of_angle: out signed(XDEG_DO_SIZ-1 downto 0)
);
end TangentUsingDegrees;

architecture RTL of TangentUsingDegrees is

--   declared constants
constant TABLE_SIZ: natural := DEG_ENTRIES*DEG_SUBENTRIES;
constant ROM_ARRAY: xdeg_da(0 to TABLE_SIZ-1) :=
    init_tbl_deg(DEG_ENTRIES,DEG_SUBENTRIES,TANF);
constant ROM_ADR_SIZ: natural := natural(ceil(log2(real(TABLE_SIZ))));
--   declared signals
signal degrees: unsigned(DEG_DI_SIZ-1 downto 0) := (others=>'0');
signal rom_adr: unsigned(DEG_DI_SIZ-1 downto 0) := (others=>'0');
signal norm_rom_adr: unsigned(ROM_ADR_SIZ-1 downto 0) := (others=>'0');
signal rom_out: signed(XDEG_TBL_WIDTH-1 downto 0) := (others=>'0');
--   state machine enumeration list
```

```
type sm_def is
(
    RESET,START_CNVRT,CHECK_BOUNDS,INDEX_TABLE,
    WAIT_ONE_CLK,CAPTURE_OUTPUT,CONVERT_SIGN
);
signal state, ret_state: sm_def := RESET;

begin
----------------------------------------------------
--   Tangent Function state machine (using degrees)
----------------------------------------------------
process(rst,clk)
begin
    if(rst='1') then
        -- outputs
        rdy <= '0';err <= '0';ovr <= '0';udf <= '0';
        func_of_angle <= (others=>'0');
        -- local
        rom_adr <= (others=>'0');
        degrees <= (others=>'0');
        -- states
        state <= RESET;
        ret_state <= RESET;
    elsif rising_edge(clk) then
        --   state machine body
        case state is
            -- reset state
            when RESET =>
                state <= START_CNVRT;
            -- wait for signal to start conversion
            when START_CNVRT =>
                if(cnvrt = '1') then
                    rdy <= '0';err <= '0';
                    ovr <= '0';udf <= '0';
                    degrees <= theta;
                    state <= CHECK_BOUNDS;
                end if;
            -- check basic bounds and directly assign boundaries of quadrants.
            -- theta of 0, 180, and 360 are handled indirectly.
            when CHECK_BOUNDS =>
                if(degrees > \360 DEGS\) then
                    rdy <= '1';
                    err <= '1';
                    state <= START_CNVRT;
```

```vhdl
                    elsif(degrees = \270 DEGS\) then
                        func_of_angle <= (others=>'0'); -- zero
                        rdy <= '1';udf <= '1';
                        state <= START_CNVRT;
                    elsif(degrees = \90 DEGS\) then
                        func_of_angle <= (others=>'0'); -- zero
                        rdy <= '1';udf <= '1';
                        state <= START_CNVRT;
                    else
                        state <= INDEX_TABLE;
                    end if;
                -- create index into ROM
                when INDEX_TABLE =>
                    if(degrees > \270 DEGS\) then
                        rom_adr <= \90 DEGS\ - (degrees - \270 DEGS\);
                        ret_state <= CONVERT_SIGN;
                    elsif(degrees > \180 DEGS\) then
                        rom_adr <= degrees - \180 DEGS\;
                        ret_state <= CAPTURE_OUTPUT;
                    elsif(degrees > \90 DEGS\) then
                        rom_adr <= \90 DEGS\ - (degrees - \90 DEGS\);
                        ret_state <= CONVERT_SIGN;
                    else
                        rom_adr <= degrees;
                        ret_state <= CAPTURE_OUTPUT;
                    end if;
                    state <= WAIT_ONE_CLK;
                when WAIT_ONE_CLK =>
                    state <= ret_state;
                when CAPTURE_OUTPUT =>
                    func_of_angle <= rom_out(rom_out'high-1 downto 0);
                    ovr <= rom_out(rom_out'high);
                    rdy <= '1';
                    state <= START_CNVRT;
                when CONVERT_SIGN =>
                    func_of_angle <= (not rom_out(rom_out'high-1 downto 0)) + 1;
                    ovr <= rom_out(rom_out'high);
                    rdy <= '1';
                    state <= START_CNVRT;
                when others =>
                    state <= RESET;
            end case;
        end if;
end process;
```

```vhdl
-- normalize rom address
norm_rom_adr <= rom_adr(norm_rom_adr'high downto 0);
-- synchronous ROM
process(clk)
begin
    if rising_edge(clk) then
        rom_out <= ROM_ARRAY(TO_INTEGER(norm_rom_adr));
    end if;
end process;

end RTL;
```

```
-------------------------------------------------------------------------
--   TangentUsingRadians.vhd (radians as input)
-------------------------------------------------------------------------
library IEEE;
use IEEE.std_logic_1164.all;
use IEEE.numeric_std.all;
use IEEE.math_real.all;
-- user packages
use work.ConversionPackageV2.all;
use work.CoreTablePackage.all;

entity TangentUsingRadians is
--   Input and output fixed-point formats are defined in CoreTablePackage.vhd
port
(
    clk,rst: in std_logic; -- system clock and reset (synchronous)
    -- inputs
    cnvrt: in std_logic; -- initiate sine function
    theta: in unsigned(RAD_DI_SIZ-1 downto 0); -- theta in radians
    -- outputs
    rdy: out std_logic; -- data ready
    err: out std_logic; -- error, input out of range
    ovr: out std_logic; -- output at its maximum range, saturated
    udf: out std_logic; -- undefined at pi/2 or 3pi/2 radians boundaries
    func_of_angle: out signed(XRAD_DO_SIZ-1 downto 0)
);
end TangentUsingRadians;

architecture RTL of TangentUsingRadians is

--   declared constants
constant TABLE_SIZ: natural := RAD_ENTRIES;
constant ROM_ARRAY: xrad_da(0 to TABLE_SIZ-1) :=
    init_tbl_rad(MATH_PI_OVER_2,TABLE_SIZ,TANF);
constant ROM_ADR_SIZ: natural := natural(ceil(log2(real(TABLE_SIZ))));
--   declared signals
signal angle: unsigned(RAD_DI_SIZ-1 downto 0) := (others=>'0');
signal rom_adr: unsigned(RAD_DI_SIZ-1 downto 0) := (others=>'0');
signal norm_rom_adr: unsigned(ROM_ADR_SIZ-1 downto 0) := (others=>'0');
signal rom_out: signed(XRAD_TBL_WIDTH-1 downto 0) := (others=>'0');
--   state machine enumeration list
type sm_def is
(
    RESET,START_CNVRT,CHECK_BOUNDS,INDEX_TABLE,
```

```
    WAIT_ONE_CLK,CAPTURE_OUTPUT,CONVERT_SIGN
);
signal state, ret_state: sm_def := RESET;

begin
-------------------------------------------------
--  Tangent Function state machine (using Radians)
-------------------------------------------------
process(rst,clk)
begin
    if(rst='1') then
        -- outputs
        rdy <= '0';err <= '0';ovr <= '0';udf <= '0';
        func_of_angle <= (others=>'0');
        -- local
        rom_adr <= (others=>'0');
        -- states
        state <= RESET;ret_state <= RESET;
    elsif rising_edge(clk) then
        --
        --   state machine body
        --
        case state is
            -- reset state
            when RESET =>
                state <= START_CNVRT;
            -- wait for signal to start conversion
            when START_CNVRT =>
                if(cnvrt = '1') then
                    rdy <= '0';err <= '0';
                    ovr <= '0';udf <= '0';
                    angle <= theta;
                    state <= CHECK_BOUNDS;
                end if;
            -- check basic bounds and directly assign boundaries of pi.
            -- angles of 0, pi, and 2pi are handled indirectly.
            when CHECK_BOUNDS =>
                if(angle > \2pi\) then
                    rdy <= '1';err <= '1';
                    state <= START_CNVRT;
                elsif(angle = \3pi/2\) then
                    func_of_angle <=  (others=>'0'); -- zero
                    rdy <= '1';udf <= '1';
                    state <= START_CNVRT;
```

```vhdl
                elsif(angle = \pi/2\) then
                    func_of_angle <=  (others=>'0'); -- zero
                    rdy <= '1';udf <= '1';
                    state <= START_CNVRT;
                else
                    state <= INDEX_TABLE;
                end if;
            -- created index into ROM
            when INDEX_TABLE =>
                if(angle > \3pi/2\) then
                    rom_adr <= \pi/2\ - (angle - \3pi/2\);
                    ret_state <= CONVERT_SIGN;
                elsif(angle > \pi\) then
                    rom_adr <= angle - \pi\;
                    ret_state <= CAPTURE_OUTPUT;
                elsif(angle > \pi/2\) then
                    rom_adr <= \pi/2\ - (angle - \pi/2\);
                    ret_state <= CONVERT_SIGN;
                else
                    rom_adr <= angle;
                    ret_state <= CAPTURE_OUTPUT;
                end if;
                state <= WAIT_ONE_CLK;
            when WAIT_ONE_CLK =>
                state <= ret_state;
            when CAPTURE_OUTPUT =>
                func_of_angle <= rom_out(rom_out'high-1 downto 0);
                ovr <= rom_out(rom_out'high);
                rdy <= '1';
                state <= START_CNVRT;
            when CONVERT_SIGN =>
                func_of_angle <= (not rom_out(rom_out'high-1 downto 0)) + 1;
                ovr <= rom_out(rom_out'high);
                rdy <= '1';
                state <= START_CNVRT;
            when others =>
                state <= RESET;
        end case;
    end if;
end process;
-- normalize rom address
norm_rom_adr <= rom_adr(norm_rom_adr'high downto 0);
```

```
-- synchronous ROM
process(clk)
begin
    if rising_edge(clk) then
        rom_out <= ROM_ARRAY(TO_INTEGER(norm_rom_adr));
    end if;
end process;

end RTL;
```

3.1.2.4   Cosecant Function (source code)

Note: *Only differences are shown, all other source code is identical to the corresponding tangent functions, both degree and radians based.*

```
--------------------------------------------------------------------
--   CosecantUsingDegrees.vhd (degrees as input)
--------------------------------------------------------------------
.
entity CosecantUsingDegrees is
.
end CosecantUsingDegrees;

architecture RTL of CosecantUsingDegrees is

--   declared constants
.
constant ROM_ARRAY: xdeg_da(0 to TABLE_SIZ-1) :=
    init_tbl_deg(DEG_ENTRIES,DEG_SUBENTRIES,CSCF);
.
begin
-------------------------------------------------
--   Cosecant Function state machine (using degrees)
-------------------------------------------------
process(rst,clk)
begin
    if(rst='1') then
        .
    elsif rising_edge(clk) then
        --   state machine body
        case state is
            .
            -- check basic bounds and directly assign boundaries of quadrants,
            -- no boundaries are handled indirectly.
            when CHECK_BOUNDS =>
                if(degrees > \360 DEGS\) then
                    rdy <= '1';
                    err <= '1';
                    state <= START_CNVRT;
                elsif(degrees = \360 DEGS\) then
                    func_of_angle <= (others=>'0'); -- zero
                    rdy <= '1';udf <= '1';
                    state <= START_CNVRT;
```

```
            elsif(degrees = \270 DEGS\) then
                func_of_angle <= XDEG_MINUS1; -- -1
                rdy <= '1';
                state <= START_CNVRT;
            elsif(degrees = \180 DEGS\) then
                func_of_angle <= (others=>'0'); -- zero
                rdy <= '1';udf <= '1';
                state <= START_CNVRT;
            elsif(degrees = \90 DEGS\) then
                func_of_angle <= XDEG_PLUS1; -- +1
                rdy <= '1';
                state <= START_CNVRT;
            elsif(degrees = \0 DEGS\) then
                func_of_angle <= (others=>'0'); -- zero
                rdy <= '1';udf <= '1';
                state <= START_CNVRT;
            else
                state <= INDEX_TABLE;
            end if;
        -- create index into ROM
        when INDEX_TABLE =>
            if(degrees > \270 DEGS\) then
                rom_adr <= \90 DEGS\ - (degrees - \270 DEGS\);
                ret_state <= CONVERT_SIGN;
            elsif(degrees > \180 DEGS\) then
                rom_adr <= degrees - \180 DEGS\;
                ret_state <= CONVERT_SIGN;
            elsif(degrees > \90 DEGS\) then
                rom_adr <= \90 DEGS\ - (degrees - \90 DEGS\);
                ret_state <= CAPTURE_OUTPUT;
            else
                rom_adr <= degrees;
                ret_state <= CAPTURE_OUTPUT;
            end if;
            state <= WAIT_ONE_CLK;

    end case;
  end if;
end process;

end RTL;
```

```
-----------------------------------------------------------------------
--   CosecantUsingRadians.vhd (radians as input)
-----------------------------------------------------------------------

.
entity CosecantUsingRadians is
.
end CosecantUsingRadians;

architecture RTL of CosecantUsingRadians is

--   declared constants
.
constant ROM_ARRAY: xrad_da(0 to TABLE_SIZ-1) :=
    init_tbl_rad(MATH_PI_OVER_2,TABLE_SIZ,CSCF);
.
begin
--------------------------------------------------
--   Cosecant Function state machine (using Radians)
--------------------------------------------------
process(rst,clk)
begin
    if(rst='1') then
                .
    elsif rising_edge(clk) then
        --
        --   state machine body
        --
        case state is
                .
            -- check basic bounds and directly assign boundaries of pi,
            -- no boundaries are handled indirectly.
            when CHECK_BOUNDS =>
                if(angle > \2pi\) then
                    rdy <= '1';err <= '1';
                    state <= START_CNVRT;
                elsif(angle = \2pi\) then
                    func_of_angle <=  (others=>'0'); -- zero
                    rdy <= '1';udf <= '1';
                    state <= START_CNVRT;
                elsif(angle = \3pi/2\) then
                    func_of_angle <= XRAD_MINUS1; -- -1
                    rdy <= '1';
                    state <= START_CNVRT;
                elsif(angle = \pi\) then
```

```
            func_of_angle <=  (others=>'0'); -- zero
            rdy <= '1';udf <= '1';
            state <= START_CNVRT;
        elsif(angle = \pi/2\) then
            func_of_angle <= XRAD_PLUS1; -- +1
            rdy <= '1';
            state <= START_CNVRT;
        elsif(angle = \0\) then
            func_of_angle <=  (others=>'0'); -- zero
            rdy <= '1';udf <= '1';
            state <= START_CNVRT;
        else
            state <= INDEX_TABLE;
        end if;
    -- created index into ROM
    when INDEX_TABLE =>
        if(angle > \3pi/2\) then
            rom_adr <= \pi/2\ - (angle - \3pi/2\);
            ret_state <= CONVERT_SIGN;
        elsif(angle > \pi\) then
            rom_adr <= angle - \pi\;
            ret_state <= CONVERT_SIGN;
        elsif(angle > \pi/2\) then
            rom_adr <= \pi/2\ - (angle - \pi/2\);
            ret_state <= CAPTURE_OUTPUT;
        else
            rom_adr <= angle;
            ret_state <= CAPTURE_OUTPUT;
        end if;
        state <= WAIT_ONE_CLK;

        .
    end case;
  end if;
end process;

.
end RTL;
```

3.1.2.5   Secant Function (source code)

Note: *Only differences are shown, all other source code is identical to the corresponding tangent functions, both degree and radians based.*

```
--------------------------------------------------------------------
--  SecantUsingDegrees.vhd     (degrees as input)
--------------------------------------------------------------------
.
entity SecantUsingDegrees is
.
end SecantUsingDegrees;

architecture RTL of SecantUsingDegrees is

--  declared constants
.
constant ROM_ARRAY: xdeg_da(0 to TABLE_SIZ-1) :=
    init_tbl_deg(DEG_ENTRIES,DEG_SUBENTRIES,SECF);
.
begin
-----------------------------------------------------
--  Secant Function state machine (using degrees)
-----------------------------------------------------
process(rst,clk)
begin
    if(rst='1') then
    .
    elsif rising_edge(clk) then
        --  state machine body
        case state is
        .
            -- create index into ROM
            when INDEX_TABLE =>
                if(degrees > \270 DEGS\) then
                    rom_adr <= \90 DEGS\ - (degrees - \270 DEGS\);
                    ret_state <= CAPTURE_OUTPUT;

                elsif(degrees > \180 DEGS\) then
                    rom_adr <= degrees - \180 DEGS\;
                    ret_state <= CONVERT_SIGN;
                elsif(degrees > \90 DEGS\) then
                    rom_adr <= \90 DEGS\ - (degrees - \90 DEGS\);
```

```
                    ret_state <= CONVERT_SIGN;
             else
                 rom_adr <= degrees;
                 ret_state <= CAPTURE_OUTPUT;
             end if;
             state <= WAIT_ONE_CLK;

       end case;
   end if;
end process;

end RTL;
```

```
-----------------------------------------------------------------------
--   SecantUsingRadians.vhd (radians as input)
-----------------------------------------------------------------------

.

entity SecantUsingRadians is

.

end SecantUsingRadians;

architecture RTL of SecantUsingRadians is

--   declared constants

.

constant ROM_ARRAY: xrad_da(0 to TABLE_SIZ-1) :=
    init_tbl_rad(MATH_PI_OVER_2,TABLE_SIZ,SECF);

.

begin
-------------------------------------------------
--   Secant Function state machine (using Radians)
-------------------------------------------------
process(rst,clk)
begin
    if(rst='1') then

        .

    elsif rising_edge(clk) then
        --   state machine body
        case state is

            .

            -- created index into ROM
            when INDEX_TABLE =>
                if(angle > \3pi/2\) then
                    rom_adr <= \pi/2\ - (angle - \3pi/2\);
                    ret_state <= CAPTURE_OUTPUT;
                elsif(angle > \pi\) then
                    rom_adr <= angle - \pi\;
                    ret_state <= CONVERT_SIGN;
                elsif(angle > \pi/2\) then
                    rom_adr <= \pi/2\ - (angle - \pi/2\);
                    ret_state <= CONVERT_SIGN;
                else
                    rom_adr <= angle;
                    ret_state <= CAPTURE_OUTPUT;
                end if;
                state <= WAIT_ONE_CLK;

                .
```

```
        end case;
    end if;
end process;

end RTL;
```

3.1.2.6   Cotangent Function (source code)

Note: *Only differences are shown, all other source code is identical to the corresponding tangent functions, both degree and radians based.*

```
--------------------------------------------------------------------------------
--   CotangentUsingDegrees.vhd      (degrees as input)
--------------------------------------------------------------------------------

.
entity CotangentUsingDegrees is

.
end CotangentUsingDegrees;

architecture RTL of CotangentUsingDegrees is

--   declared constants
.
constant ROM_ARRAY: xdeg_da(0 to TABLE_SIZ-1) :=
    init_tbl_deg(DEG_ENTRIES,DEG_SUBENTRIES,COTF);

.
begin
-------------------------------------------------
--   Cotangent Function state machine (using degrees)
-------------------------------------------------
process(rst,clk)
begin
    if(rst='1') then

    .
    elsif rising_edge(clk) then
        --   state machine body
        case state is

            .
            -- check basic bounds and directly assign boundaries of quadrants,
            -- no boundaries are handled indirectly.
            when CHECK_BOUNDS =>
                if(degrees > \360 DEGS\) then
                    rdy <= '1';
                    err <= '1';
                    state <= START_CNVRT;
                elsif(degrees = \360 DEGS\) then
                    func_of_angle <= (others=>'0'); -- zero
                    rdy <= '1';udf <= '1';
                    state <= START_CNVRT;
```

```vhdl
            elsif(degrees = \270 DEGS\) then
                func_of_angle <= (others=>'0'); -- zero
                rdy <= '1';
                state <= START_CNVRT;
            elsif(degrees = \180 DEGS\) then
                func_of_angle <= (others=>'0'); -- zero
                rdy <= '1';udf <= '1';
                state <= START_CNVRT;
            elsif(degrees = \90 DEGS\) then
                func_of_angle <= (others=>'0'); -- zero
                rdy <= '1';
                state <= START_CNVRT;
            elsif(degrees = \0 DEGS\) then
                func_of_angle <= (others=>'0'); -- zero
                rdy <= '1';udf <= '1';
                state <= START_CNVRT;
            else
                state <= INDEX_TABLE;
            end if;

        end case;
    end if;
end process;

end RTL;
```

```
-------------------------------------------------------------------------
--  CotangentUsingRadians.vhd (radians as input)
-------------------------------------------------------------------------

.

entity CotangentUsingRadians is

.

end CotangentUsingRadians;

architecture RTL of CotangentUsingRadians is

--  declared constants

.

constant ROM_ARRAY: xrad_da(0 to TABLE_SIZ-1) :=
    init_tbl_rad(MATH_PI_OVER_2,TABLE_SIZ,COTF);

.

begin
--------------------------------------------------------
--  Cotangent Function state machine (using Radians)
--------------------------------------------------------
process(rst,clk)
begin
    if(rst='1') then

        .

    elsif rising_edge(clk) then
        --
        --  state machine body
        --
        case state is

            .

            -- check basic bounds and directly assign boundaries of pi,
            -- no boundaries are handled indirectly.
            when CHECK_BOUNDS =>
                if(angle > \2pi\) then
                    rdy <= '1';err <= '1';
                    state <= START_CNVRT;
                elsif(angle = \2pi\) then
                    func_of_angle <= (others=>'0'); -- zero
                    rdy <= '1';udf <= '1';
                    state <= START_CNVRT;
                elsif(angle = \3pi/2\) then
                    func_of_angle <= (others=>'0'); -- zero
                    rdy <= '1';
                    state <= START_CNVRT;
                elsif(angle = \pi\) then
```

```
                    func_of_angle <= (others=>'0'); -- zero
                    rdy <= '1';udf <= '1';
                    state <= START_CNVRT;
                elsif(angle = \pi/2\) then
                    func_of_angle <= (others=>'0'); -- zero
                    rdy <= '1';
                    state <= START_CNVRT;
                elsif(angle = \0\) then
                    func_of_angle <= (others=>'0'); -- zero
                    rdy <= '1';udf <= '1';
                    state <= START_CNVRT;
                else
                    state <= INDEX_TABLE;
                end if;


        end case;
    end if;
end process;

end RTL;
```

3.2 Inverse Functions

Inverse functions include arcsine, arccosine, arctangent, arccosecant, arcsecant, and arccotangent. An implementation as a state machine is provided for each function in both degrees and radians. These six functions share many common parameters, which are contained in a file named *InverseTablePackage.vhd*.

Note: *a different package file is used for core functions.*

Each inverse function's input is its counterpart's (core) function-of-the-angle; whereas each inverse function output is the angle itself. This latter point is limited because inverse functions are restricted to evaluate principle values only, and within a limited domain. Core functions are multi-valued and can span all quadrants. The tables below show the relative data formatting for both input and output between core and inverse functions. As the reader can see, they do not exactly match. Criteria follows.

Core Functions (Degree Based)			
Input		**Output**	
Format	**Range**	**Format**	**Range**
UQ9.5	0 to 360°	Q1.21	+/- 1.0
		Q6.21	+/- 63.999999
Inverse Functions (Degree Based)			
Input		**Output**	
Format	**Range**	**Format**	**Range**
Q1.13	+/-1.0	Q8.8	-90 to +180°
Q2.12	+/- 3.9998		Note(1)

Core Functions (Radians Based)			
Input		**Output**	
Format	**Range**	**Format**	**Range**
UQ3.13	0 to 6.2832	Q1.27	+/- 1.0
		Q6.27	+/- 63.99999999
Inverse Functions (Radians Based)			
Input		**Output**	
Format	**Range**	**Format**	**Range**
Q1.13	+/- 1.0	Q2.16	-1.5708 to+3.1416
Q2.12	+/- 3.9998		Note(1)

Note(1): *functions with asymptotic boundaries are limited to a specific amount within the domain boundary in the package file.*

Criteria for input/output Qm.n formatting

Having identical formatting for associated core and inverse functions would be ideal but not very pragmatic. The criteria for qualifying data formatting for inverse functions in the previous tables are as follows:

- Common block RAM sizes are 16,384-bit or 32,768-bit, corresponding to address ranges of 14 and 15-bits, respectively. This directly limits the bit width of the data input, Qm.n, into the inverse functions. Combining blocks with a greater address range is possible, but this has a significant effect on performance and some synthesis tools have trouble instantiating such large synchronous ROMs. 16K blocks are preferred.
- Output formatting is a balance of the amount block RAM, typically 1-bit per block, and the output resolution based on the address range of the input. Reducing or limiting the number of duplications within memory is the objective.
- Four of the inverse functions have asymptotic behavior at their domain boundaries. Having high resolutions within these regions would result in consuming enormous amount of memory and is therefore limited with threshold boundaries. This is done primarily by limiting the integer portion of the Qm.n input.

Caveat: *Because lower and upper boundaries exist on inputs, wasted areas of memory will result. Example Arcsine range is 0.0 to 1.0. As a result the memory between 1.0 and 1.9999 is unused. Another example is the range of arccotangent is between 1.0 and 3.9998. Everything below 1.0 is unused.*

3.2.1 Inverse Table Package File

In VHDL a typical package file contains a header or declarations of the package involving data types, constants, and functions within the body portion. The body portion contains the code of the actual functions. All are visible to designs referencing the package.

Note: *Table entries correspond to positive indices. The state machine uses the sign bit from the input to determine whether the input magnitude needs to be complemented.*

Design Parameters

Degree Based Designs (Inverse Table Functions):

Inverse Table Parameters (Degrees based)		
Function	Input Range (format)	Output Range (Q8.8)
arcsine	0 to 1.0 (UQ1.13)	0 to 90°
arccosine	0 to 1.0 (UQ1.13)	90° down to 0
arctangent	0 to 3.9998 (UQ2.12)	0 to *(< 90°)
arccosecant	1.0 to 3.9998 (UQ2.12)	90° down to *(> 0)
arcsecant	1.0 to 3.9998 (UQ2.12)	0 to *(< 90°)
arccotangent	0 to 3.9998 (UQ2.12)	90° down to *(> 0)

(*) signifies an asymptotic boundary in which the actual value is never reached. These values are limited to within 15° of said boundary, which is defined by a constant.

- Supports 16384 entries, 2^{14} for 14-bits of address into the table. Input formats unsigned UQ1.13 and UQ2.12, corresponding to 1.0 and 3.9998, respectively. The former are for arcsine and arccosine, and the latter for arctangent, arccosecant, arcsecant, and arccotangent.
- The actual table only represents one quadrant worth of data. The index into the table for other quadrants is computed by the state machine.
- Input to the state machine spans the entire domain, which is limited with inverse functions to two adjacent quadrants. Negative inputs are converted to positive indexes to access table.
- The actual output of table need only span 0 to 90°, requiring unsigned UQ7.8. For simplicity, offset computations by state machine, and maintaining bit-width between functions, all are implemented as signed Q8.8.
- Extended entries have an extra flag bit in the table to indicate the output posted is a limited value within the asymptotic range. This boundary is maintained as a constant as XDEG_ASYMP_BNDS.
- Standard degree-based constants are prefaced by DEG_, extended by XDEG_. Data arrays are deg_da and xdeg_da.

- Quadrant boundaries are -90°, 0, +90°, and +180°. These constants are used for computing offsets by the state machine.
- Additional constants are provided to verify input boundaries.

Radian Based Designs (Inverse Table Functions):

Inverse Table Parameters (Radians based)		
Function	**Input Range (format)**	**Output Range (Q2.16)**
arcsine	0 to 1.0 (UQ1.13)	0 to $\pi/2$
arccosine	0 to 1.0 (UQ1.13)	$\pi/2$ down to 0
arctangent	0 to 3.9998 (UQ2.12)	0 to *(< $\pi/2$)
arccosecant	1.0 to 3.9998 (UQ2.12)	$\pi/2$ down to *(> 0)
arcsecant	1.0 to 3.9998 (UQ2.12)	0 to *(< $\pi/2$)
arccotangent	0 to 3.9998 (UQ2.12)	$\pi/2$ down to *(> 0)

(*) signifies an asymptotic boundary in which the actual value is never reached. These values are limited to within ~0.262 radians of said boundary, which is defined by a constant.

- Supports 16384 entries, 2^{14} for 14-bits of address into the table. Input formats unsigned UQ1.13 and UQ2.12, corresponding to 1.0 and 3.9998, respectively. The former are for arcsine and arccosine, and the latter for arctangent, arccosecant, arcsecant, and arccotangent.
- The actual table only represents one quadrant worth of data. The index into the table for other quadrants is computed by the state machine.
- Input to the state machine spans the entire domain, which is limited with inverse functions to two adjacent quadrants. Negative inputs are converted to positive indexes to access table.
- The actual output of table need only span 0 to $\pi/2$, requiring unsigned UQ1.16. For simplicity, offset computations by state machine, and maintaining bit-width between functions, all are implemented as signed Q2.16.
- Extended entries have an extra flag bit in the table to indicate the output posted is a limited value within the asymptotic range. This boundary is maintained as a constant as XRAD_ASYMP_BNDS.
- Standard radian-based constants are prefaced by RAD_, extended are prefaced by XRAD_. Data arrays likewise are rad_da and xrad_da.

- Quadrant boundaries are -π/2, 0, + π/2, and π. These constants are used for computing offsets.
- Additional constants are provided to verify input boundaries.

Organization Within The Header

- The first group declares constants and data array types used in degree implementations.
- The second group declares constants and data array types used in radians implementations.
- Lastly, support functions, enumeration types used by table initialization functions, and declarations of the table initialization functions.

Table Initialization Functions (Inverse Table Functions)

There are four declarations for table initialization functions, two for degrees and two for radians. Each pair's contents are the same except one returns an initialized standard data array, and one returns an initialized extended data array.

The purpose of these functions is to create a table of a specific size, index value and format, and associated output value and format, for a specific trigonometric function over a specific range. Also included are asymptotic boundary checks and endings depending on the function. These functions are called to initialize an array during instantiation and are typically inferred in synchronous block RAM or ROM.

Unlike their core counterparts, only one quadrant from 0 to 90° or π/2 radians can be initialized with these functions, and only one quadrant. Adjacent quadrants must be computed by state machine.

Next, additional information is provided to make the source code more clear to the reader.

Degree Based Functions (Inverse Table Functions)

function init_tbl_deg(func: f_def) return deg_da;

- Only the *func* parameter denotes the inverse function used to initialize the table, arcsine or arccosine. A standard data array is returned (no flag bit).
- The address range is limited from 0 to 1.0, which represents the function of the angle, and is evaluated as an integer throughout the loop. This value is also converted to its equivalent real representation, using the input Qm.n format, then applying it to the MATH_REAL.VHD inverse function. The resulting radians data is then converted to degrees, using the output Qm.n format, and placed into the corresponding address of the array.
- The size of the array is dictated by the number of bits, excluding the sign, of the input UQm.n format. Equivalent address locations above 1.0 remain undefined.

function init_tbl_deg(func: f_def) return xdeg_da;

- Only the *func* parameter denotes the inverse function used to initialize the table, arctangent, arccosecant, arcsecant, and arccotangent. The extended data array is returned (flag bit).
- Address ranges vary depending on the inverse function. A starting (0 or 1.0) and ending index (size of extended input Qm.n), which represent the function of the angle, are computed based on the function's sub-domain. This index is evaluated as an integer within the loop and is likewise converted to its real representation. The latter is applied to the corresponding inverse function from the MATH_REAL.VHD package. The resulting radians value is then converted to degrees, checked against boundaries, and placed at the corresponding address in the data array using the output Qm.n format.
- Boundaries check has to do with setting limits on data when within the asymptotic range. Arccosecant and arccotangent use a lower limit (90° - XDEG_ASYMP_BNDS), when approaching the 0° boundary. Arctangent and arcsecant use an upper limit (XDEG_ASYMP_BNDS), when approaching the 90° boundary. The flag bit is set when the corresponding data entry is limited.

- The size of the array is dictated by the number of bits, excluding the sign, of the extended input UQm.n format. Equivalent address locations below 1.0 remain undefined in arccosecant and arcsecant.

Radians Based Functions (Inverse Table Functions)

function init_tbl_rad(func: f_def) return rad_da;

- Only the *func* parameter denotes the inverse function used to initialize the table, arcsine or arccosine. A standard data array is returned (no flag bit).
- The address range is limited from 0 to 1.0, which represents the function of the angle, and is evaluated as an integer throughout the loop. This value is also converted to its equivalent real representation, using the input Qm.n format, then applying it to the MATH_REAL.VHD inverse function. The resulting radians value, using the output Qm.n format, and placed into the corresponding address of the array.
- The size of the array is dictated by the number of bits, excluding the sign, of the input UQm.n format. Equivalent address locations above 1.0 remain undefined.

function init_tbl_rad(func: f_def) return xrad_da;

- Only the *func* denotes the inverse function used to initialize the table, arctangent, arccosecant, arcsecant, and arccotangent. The extended data array is returned (flag bit).
- Address ranges vary depending on the inverse function. A starting (0 or 1.0) and ending index (size of extended input Qm.n), which represent the function of the angle, are computed based on the functions sub-domain. This index is evaluated as an integer within the loop and is also converted to its real representation. The latter is applied to the corresponding inverse function from the MATH_REAL.VHD package. The resulting radians, checked against boundaries, and placed at the corresponding address in the data array using the output Qm.n format.
- Boundaries check has to do with setting limits on data when within the asymptotic range. Arccosecant and arccotangent use a lower limit ($\pi/2$ - XRAD_ASYMP_BNDS), when approaching the 0 radians boundary. Arctangent and arcsecant have an upper limit (XRAD_ASYMP_BNDS), when approaching the $\pi/2$ boundary. The flag bit is set when the corresponding data entry is limited.

- The size of the array is dictated by the number of bits, excluding the sign, of the extended input UQm.n format. Equivalent address locations below 1.0 remain undefined in arccosecant and arcsecant.

```
-------------------------------------------------------------
--  InverseTablePackage.vhd
-------------------------------------------------------------
library IEEE;
use IEEE.std_logic_1164.all;
use IEEE.numeric_std.all;
use IEEE.math_real.all;
use work.ConversionPackageV2.all;

--  package header
package InverseTablePackage is

-- Constants for degrees based implementation. DEG_ENTRIES correspond to the
-- number of table entries for 90 degrees. Data input is a function of the angle
-- spanning +/- 1.0 for standard and +/- 3.xx for extended, signed fixed-point,
-- which corresponds to an output of -90 to +180 degrees. Extended have upper
-- bounds limit due to asymptotic behavior.
constant DEG_DI_INT_SIZ: natural := 1; -- integer bits
constant DEG_DI_FRAC_SIZ: natural := 13; -- fractional bits
constant DEG_DI_SIZ: natural := 1 + DEG_DI_INT_SIZ + DEG_DI_FRAC_SIZ;
constant DEG_ENTRIES: natural := (2**(DEG_DI_SIZ-1)); -- exclude sign bit
-- constant for extended input
constant XDEG_DI_INT_SIZ: natural := 2; -- integer bits
constant XDEG_DI_FRAC_SIZ: natural := 12; -- fractional bits
constant XDEG_DI_SIZ: natural := 1+XDEG_DI_INT_SIZ+XDEG_DI_FRAC_SIZ;
constant XDEG_ENTRIES: natural := (2**(XDEG_DI_SIZ-1)); -- exclude sign bit
 -- signed fixed-point formatting constants for standard output data
constant DEG_DO_INT_SIZ: natural := 8; -- integer bits
constant DEG_DO_FRAC_SIZ: natural := 8; -- fractional bits
constant DEG_DO_SIZ: natural := 1 + DEG_DO_INT_SIZ + DEG_DO_FRAC_SIZ;

-- signed fixed-point formatting constants for extended output
constant XDEG_TBL_WIDTH: natural := 1+DEG_DO_SIZ; -- extra flag bit
-- constants standard input boundaries
constant DEG_PLUS1: signed(DEG_DI_SIZ-1 downto 0) :=
    RealToQmn(1.0,DEG_DI_INT_SIZ,DEG_DI_FRAC_SIZ,TZERO);
constant DEG_MINUS1: signed(DEG_DI_SIZ-1 downto 0) :=
    RealToQmn(-1.0,DEG_DI_INT_SIZ,DEG_DI_FRAC_SIZ,TZERO);
-- constants for extended input boundaries
constant XDEG_PLUS1: signed(XDEG_DI_SIZ-1 downto 0) :=
    RealToQmn(1.0,XDEG_DI_INT_SIZ,XDEG_DI_FRAC_SIZ,TZERO);
constant XDEG_MINUS1: signed(XDEG_DI_SIZ-1 downto 0) :=
    RealToQmn(-1.0,XDEG_DI_INT_SIZ,XDEG_DI_FRAC_SIZ,TZERO);
constant XDEG_PLUS_MAX: signed(XDEG_DI_SIZ-1 downto 0) :=
```

```
(XDEG_DI_SIZ-1 => '0',others=>'1');
constant XDEG_MINUS_MAX: signed(XDEG_DI_SIZ-1 downto 0) :=
    (not XDEG_PLUS_MAX) + 1;
-- constants for output quadrant boundaries
constant \0 DEGS\: signed(DEG_DO_SIZ-1 downto 0)   := (others=>'0');
constant \+90 DEGS\: signed(DEG_DO_SIZ-1 downto 0)  :=
    RealToQmn(+90.0,DEG_DO_INT_SIZ,DEG_DO_FRAC_SIZ,TZERO);
constant \-90 DEGS\: signed(DEG_DO_SIZ-1 downto 0)  :=
    RealToQmn(-90.0,DEG_DO_INT_SIZ,DEG_DO_FRAC_SIZ,TZERO);
constant \+180 DEGS\: signed(DEG_DO_SIZ-1 downto 0) :=
    RealToQmn(+180.0,DEG_DO_INT_SIZ,DEG_DO_FRAC_SIZ,TZERO);
-- constants for asymptotic boundaries for extended
constant XDEG_ASYMP_BNDS: real := 75.0; -- degrees
-- standard data array
type deg_da is array (natural range <>) of signed(DEG_DO_SIZ-1 downto 0);
-- extended data array
type xdeg_da is array (natural range <>) of signed(XDEG_TBL_WIDTH-1
    downto 0);


-- Constants for radians based implementation. RAD_ENTRIES correspond to the
-- number of table entries for pi/2 radians. Data input is a function of the angle
-- spanning +/- 1.0 for standard and +/- 3.xx for extended, signed fixed-point,
-- which corresponds to an output of -pi/2 to +pi radians.
constant RAD_DI_INT_SIZ: natural := 1; -- integer bits
constant RAD_DI_FRAC_SIZ: natural := 13; -- fractional bits
constant RAD_DI_SIZ: natural := 1 + RAD_DI_INT_SIZ + RAD_DI_FRAC_SIZ;
constant RAD_ENTRIES: natural := (2**(RAD_DI_SIZ-1)); -- exclude sign bit
-- constant for extended input
constant XRAD_DI_INT_SIZ: natural := 2; -- integer bits
constant XRAD_DI_FRAC_SIZ: natural := 12; -- fractional bits
constant XRAD_DI_SIZ: natural := 1+XRAD_DI_INT_SIZ+XRAD_DI_FRAC_SIZ;
constant XRAD_ENTRIES: natural := (2**(XRAD_DI_SIZ-1)); -- exclude sign bit
 -- signed fixed-point formatting constants for standard output data
constant RAD_DO_INT_SIZ: natural := 2; -- integer bits
constant RAD_DO_FRAC_SIZ: natural := 16; -- fractional bits
constant RAD_DO_SIZ: natural := 1 + RAD_DO_INT_SIZ + RAD_DO_FRAC_SIZ;
-- signed fixed-point formatting constants for extended output data
constant XRAD_TBL_WIDTH: natural := 1 + RAD_DO_SIZ; -- extra flag bit
-- constants standard input boundaries
constant RAD_PLUS1: signed(RAD_DI_SIZ-1 downto 0) :=
    RealToQmn(1.0,RAD_DI_INT_SIZ,RAD_DI_FRAC_SIZ,TZERO);
constant RAD_MINUS1: signed(RAD_DI_SIZ-1 downto 0) :=
    RealToQmn(-1.0,RAD_DI_INT_SIZ,RAD_DI_FRAC_SIZ,TZERO);
-- constants for extended input boundaries
```

```
constant XRAD_PLUS1: signed(XRAD_DI_SIZ-1 downto 0) :=
    RealToQmn(1.0,XRAD_DI_INT_SIZ,XRAD_DI_FRAC_SIZ,TZERO);
constant XRAD_MINUS1: signed(XRAD_DI_SIZ-1 downto 0) :=
    RealToQmn(-1.0,XRAD_DI_INT_SIZ,XRAD_DI_FRAC_SIZ,TZERO);
constant XRAD_PLUS_MAX: signed(XRAD_DI_SIZ-1 downto 0) :=
    (XRAD_DI_SIZ-1 => '0',others=>'1');
constant XRAD_MINUS_MAX: signed(XRAD_DI_SIZ-1 downto 0) :=
    (not XRAD_PLUS_MAX) + 1;
-- constants for output quadrant boundaries
constant \0\: signed(RAD_DO_SIZ-1 downto 0) := (others=>'0');
constant \+pi/2\: signed(RAD_DO_SIZ-1 downto 0) :=
    RealToQmn(+1.5708,RAD_DO_INT_SIZ,RAD_DO_FRAC_SIZ,AZERO);
constant \-pi/2\: signed(RAD_DO_SIZ-1 downto 0) :=
    RealToQmn(-1.5708,RAD_DO_INT_SIZ,RAD_DO_FRAC_SIZ,AZERO);
constant \+pi\: signed(RAD_DO_SIZ-1 downto 0) :=
    RealToQmn(+3.1416,RAD_DO_INT_SIZ,RAD_DO_FRAC_SIZ,AZERO);
-- constants for asymptotic boundaries for extended
constant XRAD_ASYMP_BNDS: real :=
    (XDEG_ASYMP_BNDS * MATH_PI)/180.0;
-- standard data array
type rad_da is array (natural range <>) of signed(RAD_DO_SIZ-1 downto 0);
-- extended data array
type xrad_da is array (natural range <>) of signed(XRAD_TBL_WIDTH-1 downto
0);


-- support functions
-- returns the maximum positive qmn number for word size
function max_pos_qmn(n: integer) return signed;


-- table initialization function types
type f_def is (ARCSINF,ARCCOSF,ARCTANF,ARCCSCF,ARCSECF,ARCCOTF);
-- inverse trig functions with reciprocals, six in all
-- initialize table functions for degree based table
function init_tbl_deg(func: f_def) return deg_da;
function init_tbl_deg(func: f_def) return xdeg_da;
-- initialize table functions for radians based table
function init_tbl_rad(func: f_def) return rad_da;
function init_tbl_rad(func: f_def) return xrad_da;

end;


--   package body
package body InverseTablePackage is
```

```
-- support functions
-- returns the maximum positive qmn number for word size
function max_pos_qmn(n: integer) return signed is
variable qmn: signed(n-1 downto 0):= (others=>'1');
begin
    qmn(qmn'high) :='0'; -- clear sign bit
    return(qmn);
end function;
--
--   Initialization functions (based on degrees)
--
--   Initialize table as single quadrant representing 0-90 degrees only. Depending
--   on the function, some entries within the table are unused. The size of the
--   table is based on the input format minus the sign bit.
--
--   This function only supports Arcsine and Arccosine, so the start and end index
--   values are known, 0.0 to 1.0 over 90 degrees.  The standard degrees data
--   array is returned, no flag bit.
function init_tbl_deg(func: f_def) return deg_da is
variable input: real := 0.0;
variable radians: real := 0.0;
variable degrees: real := 0.0;
variable da: deg_da(0 to DEG_ENTRIES-1) := (others=>(others=>'X'));
begin
    -- loop through table to the maximum input supported, which is +1.0
    for i in 0 to TO_INTEGER(DEG_PLUS1) loop
        -- convert index into input for function
        input := QmnToReal(TO_SIGNED(i,DEG_DI_SIZ),DEG_DI_INT_SIZ,
            DEG_DI_FRAC_SIZ);
        -- employ math_real trig function
        case func is
                when ARCSINF => radians := ARCSIN(input);
                when ARCCOSF => radians := ARCCOS(input);
                when others => report "Function not supported!" severity failure;
        end case;
        -- convert from radians to degrees
        degrees := radians * (180.0/ MATH_PI);
        -- place results into table at specific entry
        da(i) := RealToQmn(degrees,DEG_DO_INT_SIZ,DEG_DO_FRAC_SIZ,
            TZERO);
    end loop;

    return(da);
end function;
```

```
--  This function supports Arctangent, Arccosecant, Arcsecant, and Arccotangent.
--  Starting and ending indexes are function dependent, which are known. The
--  returned data array contains a flag bit, extended, denoted by the leading x.
--  When the flag bit is set the data at location is set to its maximum or minimum
--  value as value enters the asymptotic range.
function init_tbl_deg(func: f_def) return xdeg_da is
variable start_index,end_index: natural := 0;
variable input: real:= 0.0;
variable radians: real := 0.0;
variable degrees: real := 0.0;
variable xda: xdeg_da(0 to XDEG_ENTRIES-1) := (others=>(others=>'X'));
begin

    -- set start point based on function
    case func is
        when ARCTANF => start_index := TO_INTEGER(
            RealToQmn(0.0,XDEG_DI_INT_SIZ,XDEG_DI_FRAC_SIZ,TZERO));
        when ARCCSCF => start_index := TO_INTEGER(
            RealToQmn(1.0,XDEG_DI_INT_SIZ,XDEG_DI_FRAC_SIZ,TZERO));
        when ARCSECF => start_index := TO_INTEGER(
            RealToQmn(1.0,XDEG_DI_INT_SIZ,XDEG_DI_FRAC_SIZ,TZERO));
        when ARCCOTF => start_index := TO_INTEGER(
            RealToQmn(0.0,XDEG_DI_INT_SIZ,XDEG_DI_FRAC_SIZ,TZERO));
        when others => report "Function not supported!" severity failure;
    end case;
    -- all functions have the same end point
    end_index := TO_INTEGER(max_pos_qmn(XDEG_DI_SIZ));
    -- loop through table initializing usable area within table
    for i in start_index to end_index loop
        -- convert index into input for function
        input := QmnToReal(TO_SIGNED(i,XDEG_DI_SIZ),XDEG_DI_INT_SIZ,
            XDEG_DI_FRAC_SIZ);
        -- employ math_real trig function
        case func is
            when ARCTANF => radians := ARCTAN(input);
            when ARCCSCF => radians := ARCSIN(1.0/input);
            when ARCSECF => radians := ARCCOS(1.0/input);
            when ARCCOTF =>  if(input = 0.0)then
                                radians := MATH_PI_OVER_2;
                            else
                                radians := ARCTAN(1.0/input);
                            end if;
            when others => report "Function not supported!" severity failure;
```

```
        end case;
        -- convert from radians to degrees
        degrees := radians * (180.0/ MATH_PI);
        -- check boundaries for saturation
        if((func = ARCCOTF or func = ARCCSCF) and
            degrees <= (90.0 - XDEG_ASYMP_BNDS)) then
            -- set flag bit and assign to lower boundary value
            xda(i) := '1'&RealToQmn((90.0 - XDEG_ASYMP_BNDS),
                DEG_DO_INT_SIZ,DEG_DO_FRAC_SIZ,TZERO);
        elsif((func /= ARCCOTF and func /= ARCCSCF) and
            degrees >= XDEG_ASYMP_BNDS) then
            -- set flag bit and assign to upper boundary value
            xda(i) := '1'&RealToQmn(XDEG_ASYMP_BNDS,DEG_DO_INT_SIZ,
                DEG_DO_FRAC_SIZ,TZERO);
        else
            -- convert to Qm.n data format and place in table entry along with
            -- cleared flag bit
            xda(i) := '0'&RealToQmn(degrees,DEG_DO_INT_SIZ,
                DEG_DO_FRAC_SIZ,TZERO);
        end if;
    end loop;

    return(xda);
end function;
--
--  Initialization functions (based on radians)
--
--  Initialize table as single quadrant representing pi/2 only. Depending
--  on the function, some entries within the table are unused. The size of the
--  table is based on the input format minus the sign bit.
--
--  This function only supports Arcsine and Arccosine, so the start and end index
--  values are known, 0.0 to 1.0 over pi/2 radians. The standard radians data array
--  is returned, no flag bit.
function init_tbl_rad(func: f_def) return rad_da is
variable input: real:= 0.0;
variable radians: real := 0.0;
variable da: rad_da(0 to RAD_ENTRIES-1) := (others=>(others=>'X'));
begin
    -- loop through table to the maximum input supported, which is +1.0
    for i in 0 to TO_INTEGER(RAD_PLUS1) loop
        -- convert index into input for function
        input := QmnToReal(TO_SIGNED(i,RAD_DI_SIZ),RAD_DI_INT_SIZ,
            RAD_DI_FRAC_SIZ);
```

```vhdl
    -- employ math_real trig function
    case func is
        when ARCSINF => radians := ARCSIN(input);
        when ARCCOSF => radians := ARCCOS(input);
        when others => report "Function not supported!" severity failure;
    end case;

    -- place results into table at specific entry
    da(i) := RealToQmn(radians,RAD_DO_INT_SIZ,
        RAD_DO_FRAC_SIZ,AZERO);
    end loop;

    return(da);
end function;
-- This function supports Arctangent, Arccosecant, Arcsecant, and Arccotangent.
-- Starting ending index are function dependent, which are known. The returned
-- data array contains a flag bit, extended, denoted by the leading x. When the
-- flag bit is set the data at location is set to its maximum or minimum value
-- as value enters the asymptotic range.
function init_tbl_rad(func: f_def) return xrad_da is
variable start_index,end_index: natural := 0;
variable input: real:= 0.0;
variable radians: real := 0.0;
variable adjusted: real := 0.0;
variable xda: xrad_da(0 to XRAD_ENTRIES-1) := (others=>(others=>'X'));
begin

    -- set start point based on function
    case func is
        when ARCTANF => start_index :=
TO_INTEGER(RealToQmn(0.0,XRAD_DI_INT_SIZ,XRAD_DI_FRAC_SIZ,TZERO)
);
        when ARCCSCF => start_index := TO_INTEGER(
            RealToQmn(1.0,XRAD_DI_INT_SIZ,XRAD_DI_FRAC_SIZ,TZERO));
        when ARCSECF => start_index := TO_INTEGER(
            RealToQmn(1.0,XRAD_DI_INT_SIZ,XRAD_DI_FRAC_SIZ,TZERO));
        when ARCCOTF => start_index := TO_INTEGER(
            RealToQmn(0.0,XRAD_DI_INT_SIZ,XRAD_DI_FRAC_SIZ,TZERO));
        when others => report "Function not supported!" severity failure;
    end case;
    -- all functions have the same end point. input is 1 bit wider than table
    end_index := TO_INTEGER(max_pos_qmn(XRAD_DI_SIZ));
    -- loop through table initializing usable area within table
    for i in start_index to end_index loop
```

```
-- convert index into input for function
input := QmnToReal(TO_SIGNED(i,XRAD_DI_SIZ),XRAD_DI_INT_SIZ,
    XRAD_DI_FRAC_SIZ);
-- employ math_real trig function
case func is
    when ARCTANF => radians := ARCTAN(input);
    when ARCCSCF => radians := ARCSIN(1.0/input);
    when ARCSECF => radians := ARCCOS(1.0/input);
    when ARCCOTF =>  if(input = 0.0)then
                            radians := MATH_PI_OVER_2;
                     else
                            radians := ARCTAN(1.0/input);
                     end if;
    when others => report "Function not supported!" severity failure;
end case;

-- check boundaries for saturation
if((func = ARCCOTF or func = ARCCSCF) and
    radians <= (1.5708 - XRAD_ASYMP_BNDS)) then
    -- set flag bit and assign to lower boundary value
    xda(i) := '1'&
        RealToQmn((1.5708 -
        XRAD_ASYMP_BNDS),RAD_DO_INT_SIZ,RAD_DO_FRAC_SIZ,
        TZERO);
elsif((func /= ARCCOTF and func /= ARCCSCF) and
    radians >= XRAD_ASYMP_BNDS) then
    -- set flag bit and assign to upper boundary value
    xda(i) := '1'&
        RealToQmn(XRAD_ASYMP_BNDS,RAD_DO_INT_SIZ,
        RAD_DO_FRAC_SIZ,TZERO);
else
    -- convert to Qm.n data format and place in table entry along with
    -- cleared flag bit
    xda(i) := '0'&
        RealToQmn(radians,RAD_DO_INT_SIZ,RAD_DO_FRAC_SIZ,
        AZERO);
end if;

end loop;

return(xda);
end function;

end;
```

3.2.2 Inverse State Machines

Inverse functions have restricted domains, equivalent to two quadrants of core functions. Only one quadrant worth of table data, representing the first half, is needed because the second is identical accept for sign and direction. The second quadrant is computed by the state machine.

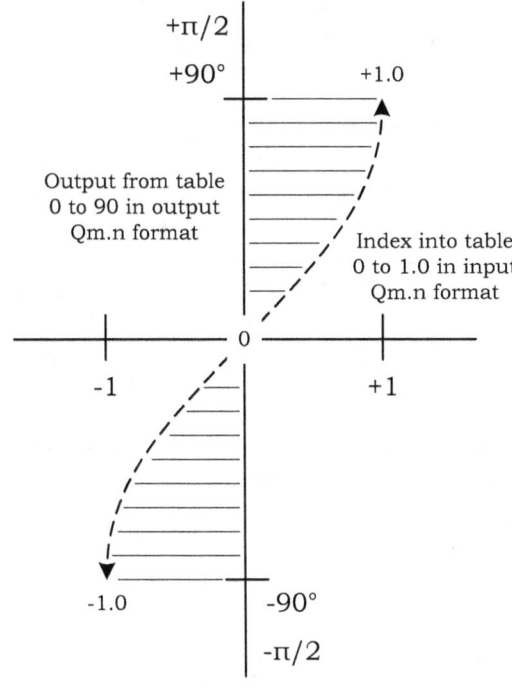

Figure 4

Graphical appearance for each inverse function is different from the other inverse functions, but they all work the same. Figure 4 above, for example, represents the arcsine function, having an input range from +/- 1.0 and a corresponding output of +/- 90° ($\pi/2$).

The number and sequence of states for inverse functions are virtually identical to core functions, Figure 3. The primary differences between core and inverse state machines is that all outputs are table driven, there are no constant values asserted at quadrant boundaries, and some outputs are adjusted by sign or offset.

Below shows the range of the table based on its usable input along with its corresponding output, for one quadrant only. *Max input* represents the maximum positive value possible with the input Qm.n format. *Boundary* represents the constant value for outputs within the asymptotic region.

If the input to the state machine is outside the input range an error is asserted; likewise when the output is at its boundary value an asymptotic signal is asserted.

Function	Table	
	Input Range	Output Range
arcsine	0 to 1.0	0 to 90°($\pi/2$)
arccosine	0 to 1.0	90°($\pi/2$) down to 0
arctangent	0 to max input	0 to boundary
arccosecant	1.0 to max input	90°($\pi/2$) down to boundary
arcsecant	1.0 to max input	0 to boundary
arccotangent	0 to max input	90°($\pi/2$) down to boundary

The state machine transfers the value of the table from input to output, as is, when the input is a positive index. Conversely, when the input is negative, it is first converted to a positive index by way of two's complement, then the table's output is arithmetically adjusted to the requirements of the second quadrant, as shown below.

Function	State Machine			
	Positive Input		Negative Input	
	Table Input	Final Output	Table Input	Final Output
arcsine	Input	Table	2'sC of Input	2'sC of Table
arccosine	Input	Table	2'sC of Input	180°(Π)–Table
arctangent	Input	Table	2'sC of Input	2'sC of Table
arccosecant	Input	Table	2'sC of Input	2'sC of Table
arcsecant	Input	Table	2'sC of Input	180°(Π)–Table
arccotangent	Input	Table	2'sC of Input	180°(Π)–Table

As with core functions the Xilinx Spartan 3E was chosen for all degree implementations and Virtex 5 for radians. Performance figures are below.

Function	Implementation	Targeted Part	Clock Speed
arcsine	Degrees	XC3S500E	142MHz
	Radians	XC5VLX30	217MHz
arccosine	Degrees	XC3S500E	139MHz
	Radians	XC5VLX30	217MHz
arctangent	Degrees	XC3S500E	140MHz
	Radians	XC5VLX30	203MHz
arccosecant	Degrees	XC3S500E	142MHz
	Radians	XC5VLX30	221MHz
arcsecant	Degrees	XC3S500E	142MHz
	Radians	XC5VLX30	223MHz
arccotangent	Degrees	XC3S500E	144MHz
	Radians	XC5VLX30	206MHz

Recap on state machine interface signals

- All degree based designs support signed Q8.8 output ranging from -90° to +180°. Input format and range depends on function. Arcsine and arccosine support signed Q1.13 input for +/-1.0; whereas arctangent, arccosecant, arcsecant, and arccotangent are extended, supporting signed Q2.12 for +/- 3.9998. Functions that check for input limits provide an *err* status bit. Extended outputs also have an *asy* status bit indicating the output is set to its asymptotic limit, which is within 15° of the boundary.

- All radians based designs support signed a Q2.16 output ranging from -π/2 to + π. Input format and range depends on function. Arcsine and arccosine support signed Q1.13 input for +/-1.0; whereas arctangent, arccosecant, arcsecant, and arccotangent are extended, supporting signed Q2.12 for +/- 3.9998. Functions that check for input limits provide an *err* status bit. Extended outputs also have an *asy* status bit indicating the output is set to its asymptotic limit, which is within 0.262 radians of the boundary.

- Synchronous ROM (using block RAM) is inferred in every design.

3.2.2.1   Arcsine Functions (source code)

```
--------------------------------------------------------------------------
-- ArcSineUsingDegrees.vhd (degrees as input)
--------------------------------------------------------------------------
library IEEE;
use IEEE.std_logic_1164.all;
use IEEE.numeric_std.all;
use IEEE.math_real.all;
-- user packages
use work.ConversionPackageV2.all;
use work.InverseTablePackage.all;

entity ArcSineUsingDegrees is
--   Input and output fixed-point formats are defined in InverseTablePackage.vhd
port
(
    clk,rst: in std_logic; -- system clock and reset (synchronous)
    -- inputs
    cnvrt: in std_logic; -- initiate sine function
    func_of_angle: in signed(DEG_DI_SIZ-1 downto 0);
    -- outputs
    rdy: out std_logic; -- data ready
    err: out std_logic; -- out of range
    degrees: out signed(DEG_DO_SIZ-1 downto 0)
);
end ArcSineUsingDegrees;

architecture RTL of ArcSineUsingDegrees is

--   declared constants
constant TABLE_SIZ: natural := DEG_ENTRIES;
constant ROM_ARRAY: deg_da(0 to TABLE_SIZ-1) := init_tbl_deg(ARCSINF);
constant ROM_ADR_SIZ: natural := natural(ceil(log2(real(TABLE_SIZ))));
--   declared signals
signal input: signed(DEG_DI_SIZ-1 downto 0) := (others=>'0');
signal rom_adr: unsigned(DEG_DI_SIZ-1 downto 0) := (others=>'0');
signal norm_rom_adr: unsigned(ROM_ADR_SIZ-1 downto 0) := (others=>'0');
signal rom_out: signed(DEG_DO_SIZ-1 downto 0) := (others=>'0');
--   state machine enumeration list
type sm_def is
(
    RESET,START_CNVRT,CHECK_BOUNDS,INDEX_TABLE,
    WAIT_ONE_CLK,CAPTURE_OUTPUT,ADJUST_OUT
```

```
);
signal state, ret_state: sm_def := RESET;

begin
-------------------------------------------------
--   ArcSine Function state machine (using degrees)
-------------------------------------------------
process(rst,clk)
begin
    if(rst='1') then
        -- outputs
        rdy <= '0';err <= '0';
        degrees <= (others=>'0');
        -- local
        rom_adr <= (others=>'0');
        input <= (others=>'0');
        -- states
        state <= RESET;
        ret_state <= RESET;
    elsif rising_edge(clk) then
        --   state machine body
        case state is
            -- reset state
            when RESET =>
                state <= START_CNVRT;
            -- wait for signal to start conversion
            when START_CNVRT =>
                if(cnvrt = '1') then
                    rdy <= '0';err <= '0';
                    input <= func_of_angle;
                    state <= CHECK_BOUNDS;
                end if;
            -- check input boundaries -1.0 to +1.0
            when CHECK_BOUNDS =>
                if(input < DEG_MINUS1 or input > DEG_PLUS1) then
                    rdy <= '1';err <= '1';
                    state <= START_CNVRT;
                else
                    state <= INDEX_TABLE;
                end if;
            -- create index into ROM
            when INDEX_TABLE =>
                -- complement index if negative, also
                -- complement the ROM's output
```

```vhdl
                if(input(input'high) = '1') then
                    rom_adr <= unsigned((not input) + 1);
                    ret_state <= ADJUST_OUT;
                else
                    rom_adr <= unsigned(input);
                    ret_state <= CAPTURE_OUTPUT;
                end if;
                state <= WAIT_ONE_CLK;
            when WAIT_ONE_CLK =>
                state <= ret_state;
            when CAPTURE_OUTPUT =>
                degrees <= rom_out;
                rdy <= '1';
                state <= START_CNVRT;
            when ADJUST_OUT =>
                degrees <= (not rom_out) + 1;
                rdy <= '1';
                state <= START_CNVRT;
            when others =>
                state <= RESET;
        end case;
    end if;
end process;
-- normalize rom address
norm_rom_adr <= rom_adr(norm_rom_adr'high downto 0);
-- synchronous ROM
process(clk)
begin
    if rising_edge(clk) then
        rom_out <= ROM_ARRAY(TO_INTEGER(norm_rom_adr));
    end if;
end process;

end RTL;
```

```
--------------------------------------------------------------------------------
--   ArcSineUsingRadians.vhd (radians as input)
--------------------------------------------------------------------------------
library IEEE;
use IEEE.std_logic_1164.all;
use IEEE.numeric_std.all;
use IEEE.math_real.all;
-- user packages
use work.ConversionPackageV2.all;
use work.InverseTablePackage.all;

entity ArcSineUsingRadians is
--   Input and output fixed-point formats are defined in InverseTablePackage.vhd
port
(
    clk,rst: in std_logic; -- system clock and reset (synchronous)
    -- inputs
    cnvrt: in std_logic; -- initiate sine function
    func_of_angle: in signed(RAD_DI_SIZ-1 downto 0); -- theta in radians
    -- outputs
    rdy: out std_logic; -- data ready
    err: out std_logic; -- out of range
    radians: out signed(RAD_DO_SIZ-1 downto 0)
);
end ArcSineUsingRadians;

architecture RTL of ArcSineUsingRadians is

--   declared constants
constant TABLE_SIZ: natural := RAD_ENTRIES;
constant ROM_ARRAY: rad_da(0 to TABLE_SIZ-1) := init_tbl_rad(ARCSINF);
constant ROM_ADR_SIZ: natural := natural(ceil(log2(real(TABLE_SIZ))));
--   declared signals
signal input: signed(RAD_DI_SIZ-1 downto 0) := (others=>'0');
signal rom_adr: unsigned(RAD_DI_SIZ-1 downto 0) := (others=>'0');
signal norm_rom_adr: unsigned(ROM_ADR_SIZ-1 downto 0) := (others=>'0');
signal rom_out: signed(RAD_DO_SIZ-1 downto 0) := (others=>'0');
--   state machine enumeration list
type sm_def is
(
    RESET,START_CNVRT,CHECK_BOUNDS,INDEX_TABLE,
    WAIT_ONE_CLK,CAPTURE_OUTPUT,ADJUST_OUT
);
signal state, ret_state: sm_def := RESET;
```

```vhdl
begin
-----------------------------------------------
--   ArcSine Function state machine (using Radians)
-----------------------------------------------
process(rst,clk)
begin
    if(rst='1') then
        -- outputs
        rdy <= '0';err <= '0';
        radians <= (others=>'0');
        -- local
        rom_adr <= (others=>'0');
        -- states
        state <= RESET;
        ret_state <= RESET;
    elsif rising_edge(clk) then
        --   state machine body
        case state is
            -- reset state
            when RESET =>
                state <= START_CNVRT;
            -- wait for signal to start conversion
            when START_CNVRT =>
                if(cnvrt = '1') then
                    rdy <= '0';err <= '0';
                    input <= func_of_angle;
                    state <= CHECK_BOUNDS;
                end if;
            -- check basic bounds and directly assign boundaries on
            -- +/- 1.0, the remaining handled indirectly
            when CHECK_BOUNDS =>
                if(input < RAD_MINUS1 or input > RAD_PLUS1) then
                    rdy <= '1';err <= '1';
                    state <= START_CNVRT;
                else
                    state <= INDEX_TABLE;
                end if;
            -- created index into ROM
            when INDEX_TABLE =>
                -- complement index if negative, also
                -- complement the ROM's output
                if(input(input'high) = '1') then
                    rom_adr <= unsigned((not input) + 1);
```

```vhdl
                ret_state <= ADJUST_OUT;
            else
                rom_adr <= unsigned(input);
                ret_state <= CAPTURE_OUTPUT;
            end if;
            state <= WAIT_ONE_CLK;
        when WAIT_ONE_CLK =>
            state <= ret_state;
        when CAPTURE_OUTPUT =>
            radians <= rom_out;
            rdy <= '1';
            state <= START_CNVRT;
        when ADJUST_OUT =>
            radians <= (not rom_out) + 1;
            rdy <= '1';
            state <= START_CNVRT;
        when others =>
            state <= RESET;
        end case;
    end if;
end process;
-- normalize rom address
norm_rom_adr <= rom_adr(norm_rom_adr'high downto 0);
-- synchronous ROM
process(clk)
begin
    if rising_edge(clk) then
        rom_out <= ROM_ARRAY(TO_INTEGER(norm_rom_adr));
    end if;
end process;

end RTL;
```

3.2.2.2 Arccosine Functions (source code)

Note: *Only differences are shown, all other source code is identical to the corresponding arcsine functions, both degree and radians based.*

```
--------------------------------------------------------------------------
--   ArcCosineUsingDegrees.vhd (degrees as input)
--------------------------------------------------------------------------

.
entity ArcCosineUsingDegrees is
.
end ArcCosineUsingDegrees;

architecture RTL of ArcCosineUsingDegrees is

--   declared constants
.
constant ROM_ARRAY: deg_da(0 to TABLE_SIZ-1) := init_tbl_deg(ARCCOSF);
.
begin
-----------------------------------------------------
--   ArcCosine Function state machine (using degrees)
-----------------------------------------------------
process(rst,clk)
begin
    if(rst='1') then
        .
        elsif rising_edge(clk) then
            --   state machine body
            case state is
                .
                when ADJUST_OUT =>
                    degrees <= \+180 DEGS\ - rom_out;
                    rdy <= '1';
                    state <= START_CNVRT;
                .
            end case;
        end if;
end process;
.
end RTL;
```

```
--------------------------------------------------------------------
--   ArcCosineUsingRadians.vhd (radians as input)
--------------------------------------------------------------------

.
entity ArcCosineUsingRadians is

.
end ArcCosineUsingRadians;

architecture RTL of ArcCosineUsingRadians is

--   declared constants

.
constant ROM_ARRAY: rad_da(0 to TABLE_SIZ-1) := init_tbl_rad(ARCCOSF);

.
begin
---------------------------------------------------
--   ArcCosine Function state machine (using Radians)
---------------------------------------------------
process(rst,clk)
begin
    if(rst='1') then

    .
    elsif rising_edge(clk) then
        --   state machine body
        case state is

            .
            when ADJUST_OUT =>
                radians <= \+pi\ - rom_out;
                rdy <= '1';
                state <= START_CNVRT;

        .
        end case;
    end if;
end process;

.
end RTL;
```

3.2.2.3   Arctangent Functions (source code)

```
-------------------------------------------------------------------------------
--  ArcTangentUsingDegrees.vhd (degrees as input)
-------------------------------------------------------------------------------
library IEEE;
use IEEE.std_logic_1164.all;
use IEEE.numeric_std.all;
use IEEE.math_real.all;
-- user packages
use work.ConversionPackageV2.all;
use work.InverseTablePackage.all;

entity ArcTangentUsingDegrees is
--  Input and output fixed-point formats are defined in InverseTablePackage.vhd
port
(
    clk,rst: in std_logic; -- system clock and reset (synchronous)
    -- inputs
    cnvrt: in std_logic; -- initiate sine function
    func_of_angle: in signed(XDEG_DI_SIZ-1 downto 0);
    -- outputs
    rdy: out std_logic; -- data ready
    asy: out std_logic; -- asymptotic region
    degrees: out signed(DEG_DO_SIZ-1 downto 0)
);
end ArcTangentUsingDegrees;

architecture RTL of ArcTangentUsingDegrees is

--  declared constants
constant TABLE_SIZ: natural := XDEG_ENTRIES;
constant ROM_ARRAY: xdeg_da(0 to TABLE_SIZ-1) := init_tbl_deg(ARCTANF);
constant ROM_ADR_SIZ: natural := natural(ceil(log2(real(TABLE_SIZ))));
--  declared signals
signal input: signed(XDEG_DI_SIZ-1 downto 0) := (others=>'0');
signal rom_adr: unsigned(XDEG_DI_SIZ-1 downto 0) := (others=>'0');
signal norm_rom_adr: unsigned(ROM_ADR_SIZ-1 downto 0) := (others=>'0');
signal rom_out: signed(XDEG_TBL_WIDTH-1 downto 0) := (others=>'0');
--  state machine enumeration list
type sm_def is
(
    RESET,START_CNVRT,CHECK_BOUNDS,INDEX_TABLE,
    WAIT_ONE_CLK,CAPTURE_OUTPUT,ADJUST_OUT
```

```
);
signal state, ret_state: sm_def := RESET;

begin
-------------------------------------------------
--   ArcTangent Function state machine (using degrees)
-------------------------------------------------
process(rst,clk)
begin
    if(rst='1') then
        -- outputs
        rdy <= '0';asy <= '0';
        degrees <= (others=>'0');
        -- local
        rom_adr <= (others=>'0');
        input <= (others=>'0');
        -- states
        state <= RESET;
        ret_state <= RESET;
    elsif rising_edge(clk) then
        --   state machine body
        case state is
            -- reset state
            when RESET =>
                state <= START_CNVRT;
            -- wait for signal to start conversion
            when START_CNVRT =>
                if(cnvrt = '1') then
                    rdy <= '0';asy <= '0';
                    input <= func_of_angle;
                    state <= CHECK_BOUNDS;
                end if;
            -- no bounds, range from 0 to +/- max input
            when CHECK_BOUNDS =>
                    state <= INDEX_TABLE; -- place holder
            -- create index into ROM
            when INDEX_TABLE =>
                -- complement index if negative, also
                -- complement the ROM's output
                if(input(input'high) = '1') then
                    rom_adr <= unsigned((not input) + 1);
                    ret_state <= ADJUST_OUT;
                else
                    rom_adr <= unsigned(input);
```

```vhdl
                ret_state <= CAPTURE_OUTPUT;
            end if;
            state <= WAIT_ONE_CLK;
        when WAIT_ONE_CLK =>
            state <= ret_state;
        when CAPTURE_OUTPUT =>
            degrees <=  rom_out(rom_out'high-1 downto 0);
            asy<= rom_out(rom_out'high);
            rdy <= '1';
            state <= START_CNVRT;
        when ADJUST_OUT =>
            degrees <=  (not rom_out(rom_out'high-1 downto 0)) + 1;
            asy<= rom_out(rom_out'high);
            rdy <= '1';
            state <= START_CNVRT;
        when others =>
            state <= RESET;
        end case;
    end if;
end process;
-- normalize rom address
norm_rom_adr <= rom_adr(norm_rom_adr'high downto 0);
-- synchronous ROM
process(clk)
begin
    if rising_edge(clk) then
        rom_out <= ROM_ARRAY(TO_INTEGER(norm_rom_adr));
    end if;
end process;

end RTL;
```

```
--------------------------------------------------------------------------
--   ArcTangentUsingRadians.vhd (radians as input)
--------------------------------------------------------------------------
library IEEE;
use IEEE.std_logic_1164.all;
use IEEE.numeric_std.all;
use IEEE.math_real.all;
-- user packages
use work.ConversionPackageV2.all;
use work.InverseTablePackage.all;

entity ArcTangentUsingRadians is
--   Input and output fixed-point formats are defined in InverseTablePackage.vhd
port
(
    clk,rst: in std_logic; -- system clock and reset (synchronous)
    -- inputs
    cnvrt: in std_logic; -- initiate sine function
    func_of_angle: in signed(XRAD_DI_SIZ-1 downto 0);
    -- outputs
    rdy: out std_logic; -- data ready
    asy: out std_logic; -- asymptotic region
    radians: out signed(RAD_DO_SIZ-1 downto 0)
);
end ArcTangentUsingRadians;

architecture RTL of ArcTangentUsingRadians is

--   declared constants
constant TABLE_SIZ: natural := XRAD_ENTRIES;
constant ROM_ARRAY: xrad_da(0 to TABLE_SIZ-1) := init_tbl_rad(ARCTANF);
constant ROM_ADR_SIZ: natural := natural(ceil(log2(real(TABLE_SIZ))));
--   declared signals
signal input: signed(XRAD_DI_SIZ-1 downto 0) := (others=>'0');
signal rom_adr: unsigned(XRAD_DI_SIZ-1 downto 0) := (others=>'0');
signal norm_rom_adr: unsigned(ROM_ADR_SIZ-1 downto 0) := (others=>'0');
signal rom_out: signed(XRAD_TBL_WIDTH-1 downto 0) := (others=>'0');
--   state machine enumeration list
type sm_def is
(
    RESET,START_CNVRT,CHECK_BOUNDS,INDEX_TABLE,
    WAIT_ONE_CLK,CAPTURE_OUTPUT,ADJUST_OUT
);
signal state, ret_state: sm_def := RESET;
```

```vhdl
begin
-------------------------------------------------------
--  ArcTangent Function state machine (using radians)
-------------------------------------------------------
process(rst,clk)
begin
    if(rst='1') then
        -- outputs
        rdy <= '0';asy <= '0';
        radians <= (others=>'0');
        -- local
        rom_adr <= (others=>'0');
        input <= (others=>'0');
        -- states
        state <= RESET;
        ret_state <= RESET;
    elsif rising_edge(clk) then
        --  state machine body
        case state is
            -- reset state
            when RESET =>
                state <= START_CNVRT;
            -- wait for signal to start conversion
            when START_CNVRT =>
                if(cnvrt = '1') then
                    rdy <= '0';asy <= '0';
                    input <= func_of_angle;
                    state <= CHECK_BOUNDS;
                end if;
            -- no bounds, range from 0 to +/- max input
            when CHECK_BOUNDS =>
                    state <= INDEX_TABLE; -- place holder
            -- create index into ROM
            when INDEX_TABLE =>
                -- complement index if negative, also
                -- complement the ROM's output
                if(input(input'high) = '1') then
                    rom_adr <= unsigned((not input) + 1);
                    ret_state <= ADJUST_OUT;
                else
                    rom_adr <= unsigned(input);
                    ret_state <= CAPTURE_OUTPUT;
                end if;
```

```vhdl
                state <= WAIT_ONE_CLK;
            when WAIT_ONE_CLK =>
                state <= ret_state;
            when CAPTURE_OUTPUT =>
                radians <=  rom_out(rom_out'high-1 downto 0);
                asy<= rom_out(rom_out'high);
                rdy <= '1';
                state <= START_CNVRT;
            when ADJUST_OUT =>
                radians <=  (not rom_out(rom_out'high-1 downto 0)) + 1;
                asy<= rom_out(rom_out'high);
                rdy <= '1';
                state <= START_CNVRT;
            when others =>
                state <= RESET;
        end case;
    end if;
end process;
-- normalize rom address
norm_rom_adr <= rom_adr(norm_rom_adr'high downto 0);
-- synchronous ROM
process(clk)
begin
    if rising_edge(clk) then
        rom_out <= ROM_ARRAY(TO_INTEGER(norm_rom_adr));
    end if;
end process;

end RTL;
```

3.2.2.4   Arccosecant Functions (source code)

Note: *Only differences are shown, all other source code is identical to the corresponding arctangent functions, both degree and radians based.*

```
---------------------------------------------------------------
-- ArcCosecantUsingDegrees.vhd (degrees as input)
---------------------------------------------------------------

entity ArcCosecantUsingDegrees is
-- Input and output fixed-point formats are defined in InverseTablePackage.vhd
port
(
    .
    err: out std_logic; -- out of range
    .
);
end ArcCosecantUsingDegrees;

architecture RTL of ArcCosecantUsingDegrees is

-- declared constants
    .
constant ROM_ARRAY: xdeg_da(0 to TABLE_SIZ-1) := init_tbl_deg(ARCCSCF);
    .
begin
-----------------------------------------------------
-- ArcCosecant Function state machine (using degrees)
-----------------------------------------------------
process(rst,clk)
begin
    if(rst='1') then
        .
        rdy <= '0';err <= '0';asy <= '0';
        .
    elsif rising_edge(clk) then
        -- state machine body
        case state is
            .
            -- wait for signal to start conversion
            when START_CNVRT =>
                if(cnvrt = '1') then
                    rdy <= '0';err <= '0';asy <= '0';
```

```
                input <= func_of_angle;
                state <= CHECK_BOUNDS;
            end if;
        -- check input boundaries -max to -1 or +1 to +max
        when CHECK_BOUNDS =>
            if(input > XDEG_MINUS1 and input < XDEG_PLUS1) then
                rdy <= '1';err <= '1';
                state <= START_CNVRT;
            else
                state <= INDEX_TABLE;
            end if;

    end case;
  end if;
end process;

end RTL;
```

```
----------------------------------------------------------------
-- ArcCosecantUsingRadians.vhd (radians as input)
----------------------------------------------------------------

entity ArcCosecantUsingRadians is
-- Input and output fixed-point formats are defined in InverseTablePackage.vhd
port
(
    .
    err: out std_logic; -- out of range
    .
);
end ArcCosecantUsingRadians;

architecture RTL of ArcCosecantUsingRadians is

--   declared constants
    .
constant ROM_ARRAY: xrad_da(0 to TABLE_SIZ-1) := init_tbl_rad(ARCCSCF);
    .
begin
-----------------------------------------------------
--   ArcCosecant Function state machine (using radians)
-----------------------------------------------------
process(rst,clk)
begin
    if(rst='1') then
        .
        err <= '0';
        .
    elsif rising_edge(clk) then
        --   state machine body
        case state is
            .
            -- wait for signal to start conversion
            when START_CNVRT =>
                if(cnvrt = '1') then
                    rdy <= '0';err <= '0';asy <= '0';
                    input <= func_of_angle;
                    state <= CHECK_BOUNDS;
                end if;
            -- check input boundaries -max to -1 or +1 to +max
            when CHECK_BOUNDS =>
                if(input > XRAD_MINUS1 and input < XRAD_PLUS1) then
```

```
                        rdy <= '1';err <= '1';
                        state <= START_CNVRT;
                    else
                        state <= INDEX_TABLE;
                    end if;

            end case;
        end if;
end process;

end RTL;
```

3.2.2.5 Arcsecant Functions (source code)

Note: *Only differences are shown, all other source code is identical to the corresponding arctangent functions, both degree and radians based.*

```
----------------------------------------------------------------
-- ArcSecantUsingDegrees.vhd (degrees as input)
----------------------------------------------------------------

.
entity ArcSecantUsingDegrees is
--   Input and output fixed-point formats are defined in InverseTablePackage.vhd
port
(
    .
    err: out std_logic; -- out of range
    .
);
end ArcSecantUsingDegrees;

architecture RTL of ArcSecantUsingDegrees is

--   declared constants
.
constant ROM_ARRAY: xdeg_da(0 to TABLE_SIZ-1) := init_tbl_deg(ARCSECF);
.

begin
-------------------------------------------------
--   ArcSecant Function state machine (using degrees)
-------------------------------------------------
process(rst,clk)
begin
    if(rst='1') then
        .
        err <= '0';
        .
    elsif rising_edge(clk) then
        --
        --   state machine body
        --
        case state is
            -- reset state
            when RESET =>
```

```
                state <= START_CNVRT;
            -- wait for signal to start conversion
            when START_CNVRT =>
                if(cnvrt = '1') then
                    rdy <= '0';err <= '0';asy <= '0';
                    input <= func_of_angle;
                    state <= CHECK_BOUNDS;
                end if;
            -- check input boundaries -max to -1 or +1 to +max
            when CHECK_BOUNDS =>
                if(input > XDEG_MINUS1 and input < XDEG_PLUS1) then
                    rdy <= '1';err <= '1';
                    state <= START_CNVRT;
                else
                    state <= INDEX_TABLE;
                end if;

            when ADJUST_OUT =>
                degrees <= \+180 DEGS\ - rom_out(rom_out'high-1 downto 0);
                asy<= rom_out(rom_out'high);
                rdy <= '1';
                state <= START_CNVRT;
            when others =>
                state <= RESET;
        end case;
    end if;
end process;

end RTL;
```

```
--------------------------------------------------------------------------------
--   ArcSecantUsingRadians.vhd (radians as input)
--------------------------------------------------------------------------------

.

entity ArcSecantUsingRadians is
--   Input and output fixed-point formats are defined in InverseTablePackage.vhd
port
(

    .

    err: out std_logic; -- out of range

    .

);
end ArcSecantUsingRadians;

architecture RTL of ArcSecantUsingRadians is

--   declared constants

    .

constant ROM_ARRAY: xrad_da(0 to TABLE_SIZ-1) := init_tbl_rad(ARCSECF);

.

begin
----------------------------------------------------
--   ArcSecant Function state machine (using radians)
----------------------------------------------------
process(rst,clk)
begin
    if(rst='1') then

        .

        err <= '0';

        .

    elsif rising_edge(clk) then
        --   state machine body
        case state is

            .

            -- wait for signal to start conversion
            when START_CNVRT =>
                if(cnvrt = '1') then
                    rdy <= '0';err <= '0';asy <= '0';
                    input <= func_of_angle;
                    state <= CHECK_BOUNDS;
                end if;
            -- check input boundaries -max to -1 or +1 to +max
            when CHECK_BOUNDS =>
                if(input > XRAD_MINUS1 and input < XRAD_PLUS1) then
```

```
                rdy <= '1';err <= '1';
                state <= START_CNVRT;
            else
                state <= INDEX_TABLE;
            end if;

        when ADJUST_OUT =>
            radians <=  \+pi\ - rom_out(rom_out'high-1 downto 0);
            asy<= rom_out(rom_out'high);
            rdy <= '1';
            state <= START_CNVRT;
        when others =>
            state <= RESET;
    end case;
  end if;
end process;

end RTL;
```

3.2.2.6   Arccotangent Functions (source code)

Note: *Only differences are shown, all other source code is identical to the corresponding arctangent functions, both degree and radians based.*

```
-----------------------------------------------------------------------------
--  ArcCotangentUsingDegrees.vhd (degrees as input)
-----------------------------------------------------------------------------

.
entity ArcCotangentUsingDegrees is

.
end ArcCotangentUsingDegrees;
architecture RTL of ArcCotangentUsingDegrees is

--   declared constants

.
constant ROM_ARRAY: xdeg_da(0 to TABLE_SIZ-1) := init_tbl_deg(ARCCOTF);

.
begin
------------------------------------------------------
--  ArcCotangent Function state machine (using degrees)
------------------------------------------------------
process(rst,clk)
begin
    if(rst='1') then

        .
    elsif rising_edge(clk) then
        --   state machine body
        case state is

            .
            when ADJUST_OUT =>
                degrees <= \+180 DEGS\ - rom_out(rom_out'high-1 downto 0);
                asy<= rom_out(rom_out'high);
                rdy <= '1';
                state <= START_CNVRT;

            .
        end case;
    end if;
end process;

.
end RTL;
```

```
-------------------------------------------------------------------------
--   ArcCotangentUsingRadians.vhd (radians as input)
-------------------------------------------------------------------------

.

entity ArcCotangentUsingRadians is

.

end ArcCotangentUsingRadians;

architecture RTL of ArcCotangentUsingRadians is

--   declared constants

.

constant ROM_ARRAY: xrad_da(0 to TABLE_SIZ-1) := init_tbl_rad(ARCCOTF);

.

begin
-------------------------------------------------------
--   ArcCotangent Function state machine (using radians)
-------------------------------------------------------
process(rst,clk)
begin
    if(rst='1') then

        .

    elsif rising_edge(clk) then
        --   state machine body
        case state is

            .

            when ADJUST_OUT =>
                radians <=  \+pi\ - rom_out(rom_out'high-1 downto 0);
                asy<= rom_out(rom_out'high);
                rdy <= '1';
                state <= START_CNVRT;

        end case;
    end if;
end process;

.

end RTL;
```

4 Real-Time Solutions

The advantage of real-time solutions in this chapter is precision. Table-based designs have an inherent limitation due to the address range of the block RAM, which corresponds to the number of bits in the operand -- in most cases the fractional portion of the operand. The fractional portion of all designs in this chapter default to 32-bits and can be expanded even further.

There is a mathematical solution for all trigonometric functions utilizing the Taylor-Series. Those functions with periodicity characteristics work well, such as sine and cosine (tangent indirectly), but those that have asymptotic regions or exact domain boundaries do not. This latter point is because the number of polynomials and the increasing order required -- make such a design impractical. Moreover, convergence is very slow.

Borchardt-Gauss iteration offers faster convergence without the need for figuring increasingly complex coefficients required for a lengthy Taylor expansion. Thus is the solutions for the inverse functions arcsine, arccosine, and arctangent.

While very different algorithms, both use multiplication and division. Borchardt-Gauss also uses a square root function. These functions are not recreated here, but are external to each trigonometric function's module. The reader may reference Volumes 2, 3, and 5 of this book series for Multipliers, Dividers, and Root Functions, respectively, or design their own, or take advantage of DSP facilities available in some synthesis tool sets. Adders, however, are inferred within each state machine design and are expected to be fast because the targeted device is a Virtex-5, which has high performance embedded adders within the silicon.

Other than the external multipliers, dividers, and square root functions, each design is self-contained, requiring only a short package file for constants and data formats used throughout.

4.1 Taylor-Series

In calculus, the subject of Infinite Series is a series of terms that are summed together -- an approximation – Taylor-Series being in a special subcategory. Particulars include: terms within the series are represented by a string of polynomials, sometimes referred to as

Taylor polynomials, and the center of approximation can be any value of x, not just x as zero.

While referred to as an infinite series, infinite expressions are not realistic, because we are not modeling a function but actually creating hardware circuits to carry out the work. On the other hand, being too restrictive on the number of polynomials, or the nth degree, will result in poor approximations and limit range of the functions usefulness. This must also be balanced with the use of resources like adders, multipliers, and dividers, and of course the total execution time.

Issues like convergence and divergence must be considered as well. For example, a Taylor-Series sine function of the 9th degree can produce an accurate approximation over an input range of $\pi/2$ radians. Pi radians would require at least an 11th degree. All previous table-based designs were implemented using 1 quadrant or $\pi/2$ radians worth of data -- other quadrants were computed with that data. That approach is used here as well.

As for design structure, a single process will be used, as shown in the source code fragment below. Independent case statement bodies operate as threads, each handling a different aspect of the algorithm, and operate concurrently. Threads can also signal or be signaled by other threads, thus accounting for dependencies and handshake, as well as independently accessing external modules.

```
name: process(rst,clk)
begin
    if(rst='1') then
        -- object initialization
    elsif rising_edge(clk) then
        -- one clock signals

        -- case statement body (master thread)

        -- case statement body (thread 1)

        -- case statement body (thread 2)
        .
        -- case statement body (thread n)

    end if;
end process;
```

Three designs are provided using the Taylor Series: sine, cosine, and tangent, the latter being computed indirectly. All designs are radians based, and all target a Virtex-5 FPGA device. Each subsection contains the function's notation in sigma and expanded form, graphs, divide and conquer strategies, input/output range, and source code.

4.1.1 Common Features

Master thread

Many similarities exist between each of the three Taylor-Series designs: sine, cosine, and tangent. In particular the master thread. While some of the particulars are different, the basic sequence is the same as diagramed in Figure 5.

First, if the input is greater than 2π an error is asserted. If on a boundary of π a constant is asserted as the output data (this prevents rounding issues from interfering with exact values).

Functions	Quadrant Boundaries -- versus -- Data Output			
	0/2π	**π/2**	**π**	**3π/2**
Sine	0	+1.0	0	-1.0
Cosine	+1.0	0	-1.0	0
Tangent	0	*Undefined	0	*Undefined

(*): undefined represents the exact quadrant boundary which there is no solution. An extra signal called "udf" indicates this condition.

If the input is neither greater than 2π or on a quadrant boundary, x is computed to be within 1 quadrant, as well as determining the resulting sign. Computation of x is the same for all designs but the resulting sign is not.

Input Range (angle)	Creation of x
Greater than **0** but less than **$\pi/2$**	x = angle
Greater than **$\pi/2$** but less than **π**	$x = \pi/2$ - (angle - $\pi/2$)
Greater than **π** but less than **$3\pi/2$**	x = angle - π
Greater than **3 $\pi/2$** but less than **2π**	$x = \pi/2$- (angle - $3\pi/2$)

Function	Input Range (angle) -- versus – Resulting Sign			
	>0 to <2/π	>$\pi/2$ to <π	>π to <$3\pi/2$	>$3\pi/2$ to <2π
Sine	+	+	-	-
Cosine	+	-	-	+
Tangent	+	-	+	-

Once x is created, the subordinate threads are started.

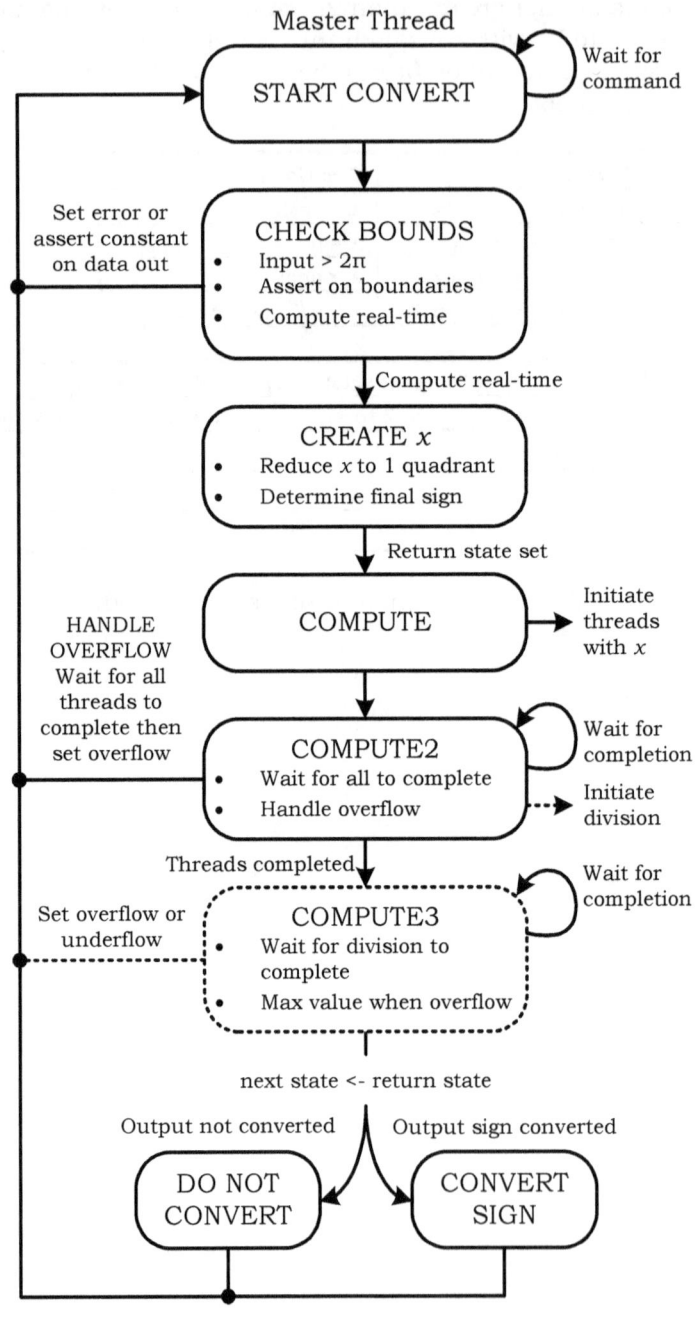

Figure 5

State COMPUTE2 in Figure 5 waits for all threads to complete their respective sequences, or if an overflow occurred during a multiplication. If the latter is true, the HANDLE OVERFLOW state is entered, waiting for all threads to return to their idle state, at which time an overflow error is asserted by the master thread. If all threads completed error free, the result is either passed as is, or is converted to a negative number, based on the return state set when x was created.

Note: *The tangent algorithm employs a division cycle as the last step before sign conversion, COMPUTE3, depicted by the dotted lines.*

External Resources

Multiplication and division are all handled by external modules, signals being available on the port pins of the entity. Concurrent operations of external modules greatly improves performance.

Functions	External Multipliers	External Dividers
Sine	2	0
Cosine	2	0
Tangent	3	1

Note: *Associated adders within each respective function directly determines the clock speed of the module (without including external multipliers/dividers).*

Caveat: *Rounding is a function of the external modules used.*

Performance

Function	Targeted Device	Clockspeed
Sine	XC5VLX30	233MHz
Cosine	XC5VLX30	222MHz
Tangent	XC5VLX110(1)	224MHz

Note1: *The logic resources required for tangent can fit within a XC5VLX30, XC5VLX50 recommended, but the IOB count is too high because of the additional multiplier and divider, which would normally be elsewhere in the same FPGA or ASIC device. The XC5VLX110 was selected because of the IOB requirement.*

Handshaking between threads

Handshaking between threads is accomplished by advancing and monitoring counters, which are associated with principle data objects. When one of these data objects is updated its corresponding counter is incremented as well. Threads use count values from their own counters and the counters of other threads to regulate coordination and progression through the algorithm. Each design will provide truth tables showing these dependencies.

Possible Improvements

The use of faster adders when combining computed polynomial terms and when converting the result to a negative numbers, via two's complement.

4.1.2 Taylor–Series Package File

All Taylor-Series functions require a short package file called *TayloRealTimePackage.vhd*.

Design Parameters

- The input data format defaults to Q3.32, supporting an input from 0 to 2π radians. The integer portion should remain at 3-bits, but the fractional portion may be increased to gain greater accuracy/precision. The master thread of the state machine reduces this input to a single quadrant of range, 0 to $\pi/2$ radians.
- Internal data paths employ Q6.32 to account for the largest value possible within the polynomial terms under normal operation, x^9 or $(\pi/2)^9$.
- Data out for sine and cosine use Q1.32 for a range of +/-1.0. Tangent, however, requires the same format as the divider employed, defaulting to Q6.32 with a range of +/-63.9999999998 (compatible with table-based designs). The divider format directly dictates the saturation value when within the asymptotes range. The divider signals the module whenever it detects an overflow when computing the quotient.

Note: *although the number of fractional bits can be increased, all other data formats within the design must have the same number of fractional bits for the resize operator to work properly.*

Organization Within the Header

- Constants defining data formats for input, internal, and output data paths.
- Constants for comparing input radians, computing x, and output constants.
- Function declaration.

Functions

- Function for returning the maximum positive number for a given data format Qm.n.

```
--------------------------------------------------------------
--   TaylorRealTimePackage.vhd
--------------------------------------------------------------
library IEEE;
use IEEE.std_logic_1164.all;
use IEEE.numeric_std.all;
use IEEE.math_real.all;
use work.ConversionPackageV2.all;


--   package header
package TaylorRealTimePackage is

    --
    --   Fixed-point data formats
    --
    -- standard data input for sine, cosine, and tangent (0 to 2pi radians). Having
    -- the same number of fractional bits for all data formats ensures the resize
    -- operator works properly.
    constant DI_INT_SIZ: natural := 3; -- integer bits
    constant DI_FRAC_SIZ: natural := 32; -- fractional bits
    constant DI_SIZ: natural := 1 + DI_INT_SIZ + DI_FRAC_SIZ;
    -- standard data output for sine and cosine
    constant DO_INT_SIZ: natural := 1; -- integer bits
    constant DO_FRAC_SIZ: natural := DI_FRAC_SIZ; -- fractional bits
    constant DO_SIZ: natural := 1 + DO_INT_SIZ + DO_FRAC_SIZ;

    -- internal multiplication data formats: multiplier, multiplicand,
    -- and product, plus data objects.
    constant M_INT_SIZ: natural := 6; -- integer bits
    constant M_FRAC_SIZ: natural := DI_FRAC_SIZ; -- fractional bits
    constant M_SIZ: natural := 1 + M_INT_SIZ + M_FRAC_SIZ;
    -- internal division data formats: divisor, dividend, and quotient,
```

```vhdl
-- plus data objects, and tangent output
constant D_INT_SIZ: natural := 6; -- integer bits
constant D_FRAC_SIZ: natural := DI_FRAC_SIZ; -- fractional bits
constant D_SIZ: natural := 1 + D_INT_SIZ + D_FRAC_SIZ;
--
--   Data constants
--
-- constants for input quadrant boundaries
constant \0\: signed(DI_SIZ-1 downto 0) := (others=>'0');
constant \pi/2\: signed(DI_SIZ-1 downto 0) :=
    RealToQmn(1.5708,DI_INT_SIZ,DI_FRAC_SIZ,AZERO);
constant \pi\: signed(DI_SIZ-1 downto 0) :=
    RealToQmn(3.1416,DI_INT_SIZ,DI_FRAC_SIZ,AZERO);
constant \3pi/2\: signed(DI_SIZ-1 downto 0) :=
    RealToQmn(4.7124,DI_INT_SIZ,DI_FRAC_SIZ,AZERO);
constant \2pi\: signed(DI_SIZ-1 downto 0) :=
    RealToQmn(6.2832,DI_INT_SIZ,DI_FRAC_SIZ,AZERO);
-- constants for standard output boundaries
constant PLUS1: signed(DO_SIZ-1 downto 0) :=
    RealToQmn(1.0,DO_INT_SIZ,DO_FRAC_SIZ,TZERO);
constant MINUS1: signed(DO_SIZ-1 downto 0) :=
    RealToQmn(-1.0,DO_INT_SIZ,DO_FRAC_SIZ,TZERO);
--
-- support functions
--
-- returns the maximum positive qmn number for word size
function max_pos_qmn(n: integer) return signed;

end;

--   package body
package body TaylorRealTimePackage is

    -- support functions
    -- returns the maximum positive qmn number for word size
    function max_pos_qmn(n: integer) return signed is
    variable qmn: signed(n-1 downto 0):= (others=>'1');
    begin
        qmn(qmn'high) :='0'; -- clear sign bit
        return(qmn);
    end function;

end;
```

4.1.3 Sine Function (Taylor-Series)

Figure 6 below shows the Taylor-Series sine function in its sigma and expanded notation, followed by its modeled graph. The solid line was created with an accurate software algorithm. The dashed line represents the simulated Taylor-Series expression of the 9th degree over π radians.

$$\sum_{n=0}^{4} (-1)^n \frac{x^{2n+1}}{(2n+1)!}$$

$$\sin x = x - \frac{x^3}{3!} + \frac{x^5}{5!} - \frac{x^7}{7!} + \frac{x^9}{9!}$$

$$(3 \cdot 2 \cdot 1)$$
$$(5 \cdot 4 \cdot 3 \cdot 2 \cdot 1)$$
$$(7 \cdot 6 \cdot 5 \cdot 4 \cdot 3 \cdot 2 \cdot 1)$$
$$(9 \cdot 8 \cdot 7 \cdot 6 \cdot 5 \cdot 4 \cdot 3 \cdot 2 \cdot 1)$$

Coefficients from factorials

Figure 6

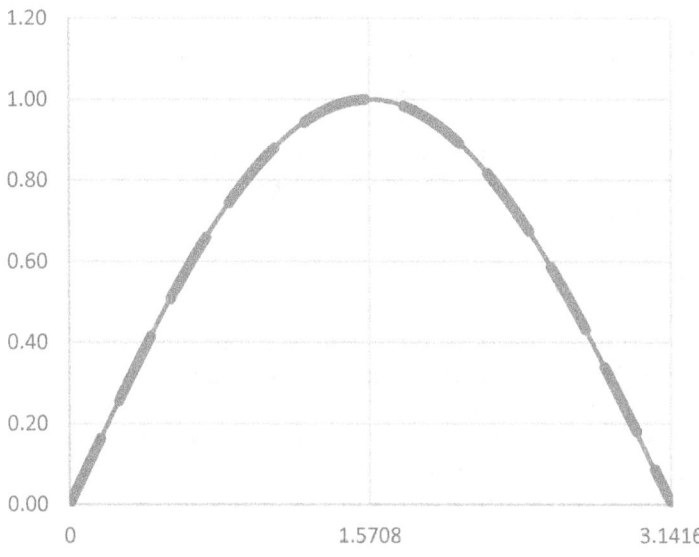

While barely perceivable, divergence begins to occur slightly before the π radians boundary of 3.1416, which is in the lower right-hand corner of the graph. This means that an expansion of the 9th degree is only suitable for $\pi/2$ radians, 1.5708, which is one quadrant of range and just what we want. Only a single quadrant of range is required, the state machine will compute the other quadrants.

Note: *An expansion of 7th degree would be inaccurate when approaching the $\pi/2$ boundary, making it unsuitable for one quadrant.*

Threads

In addition to the master thread, which checks boundaries and computes the value of x to be operated on, several subordinate threads are required. By examining the expanded expression in Figure 6, work can logically be partitioned into three parts or three threads.

- Thread 1 – increases the value of x to is nth power at each stage of the algorithm, Δx (delta x). Requiring a dedicated external multiplier.

- Thread 2 – computes each polynomial term at each stage of the algorithm, Δt (delta term). Requiring a dedicated external multiplier as well.

- Thread 3 – computes the collective sum as each polynomial term is retired, Δs (delta sum). Uses local adder.

Each thread runs independently and concurrently with all other threads. However, threads must coordinate their flow with the flow of other threads.

Handshake

Counters are used to indicate the status of each thread, one counter per thread. Each time its associated data object is updated with the results of the most recent operation, Δx, Δt, or Δs, the counter is advanced. Conversely, before the next operation on a data object begins, certain conditions must be satisfied, namely counters and the occurrence of other control signals.

Note: *Data objects Δx, Δt, and Δs are initialized to x, 0, and x.*

Truth table for controlling Thread 1, computing the power of x.

Dependencies			Action/Operation
Δx cnt	Δt cnt	New Δt started	
0,1,3,5,7	X	X	
2	0	Yes	$\Delta x = \Delta x \bullet x$, then increment Δx cnt
4	1	Yes	
6	2	Yes	

Truth table for controlling Thread 2, computing polynomial terms.

Dependencies		Action/Operation
Δx cnt	Δt cnt	
2	0	
4	1	$\Delta t = $ coefficient $\bullet \Delta x$, then increment Δt cnt
6	2	
8	3	

Note: *Coefficients are supplied by a constant array, indexed by the current Δt count.*

Truth table for controlling Thread 3, summing polynomial terms.

Dependencies		Action/Operation
Δt cnt	Δs cnt	
1	0	$\Delta s = \Delta s - \Delta t$, then increment Δs cnt
2	1	$\Delta s = \Delta s + \Delta t$, then increment Δs cnt
3	2	$\Delta s = \Delta s - \Delta t$, then increment Δs cnt
4	3	$\Delta s = \Delta s + \Delta t$, then increment Δs cnt

Figure 7 represents the progression of each thread in accordance with the truth tables above. Once all counts reach their maximum count value of 8, 4, and 4, sin x has been computed, at which time the master thread completes the final operation of assigning the proper sign to the result and then asserting the *rdy* signal.

Taylor-Series Sine Function Progression

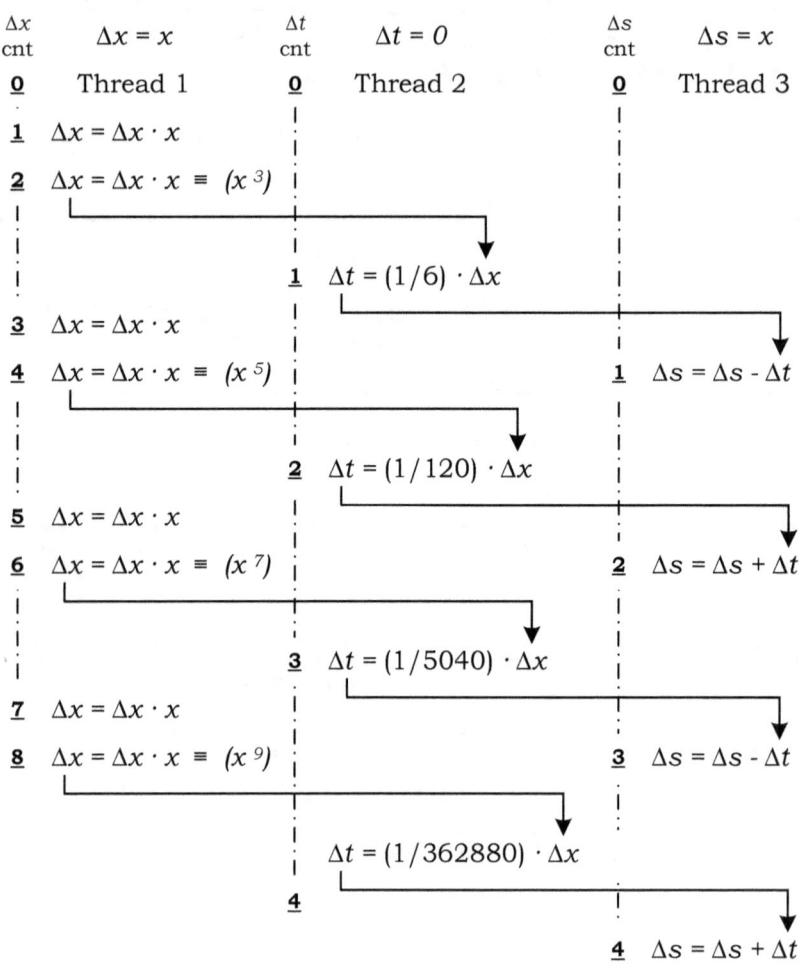

Figure 7

Recap on state machine design for Sine function (Taylor-Series)

- Design support Q3.32 for a range of 0 to 2π radians input, and Q1.32 for an output ranging from -1.0 to +1.0
- An *err* status bit indicates the input has exceeded 2π radians, an *ovrflw* bit indicates an overflow occurred in one of the external multipliers and the result is invalid.

4.1.3.1 Sine Function (source code)

```
-------------------------------------------------------------------------------
--  SineUsingTaylor.vhd (radians as input)
-------------------------------------------------------------------------------
library IEEE;
use IEEE.std_logic_1164.all;
use IEEE.numeric_std.all;
use IEEE.math_real.all;
-- user packages
use work.ConversionPackageV2.all;
use work.TaylorRealTimePackage.all;

entity SineUsingTaylor is
--  Input and output fixed-point formats, as well as internal data paths,
--  are defined in TaylorRealTimePackage.vhd
port
(
    clk,rst: in std_logic; -- system clock and reset (synchronous)
    -- inputs
    cnvrt: in std_logic; -- initiate sine function
    theta: in signed(DI_SIZ-1 downto 0); -- theta in radians
    -- interface to external multiplier 1
    start1: inout std_logic; -- start multiplication
    multiplier1: out signed(M_SIZ-1 downto 0);
    multiplicand1: out signed(M_SIZ-1 downto 0);
    cmplt1: in std_logic; -- multiplication complete
    ovrflw1: in std_logic; -- overflow error
    product1: in signed(M_SIZ-1 downto 0);
    -- interface to external multiplier 2
    start2: inout std_logic; -- start multiplication
    multiplier2: out signed(M_SIZ-1 downto 0);
    multiplicand2: out signed(M_SIZ-1 downto 0);
    cmplt2: in std_logic; -- multiplication complete
    ovrflw2: in std_logic; -- overflow error
    product2: in signed(M_SIZ-1 downto 0);
```

```vhdl
    -- outputs
    rdy: out std_logic; -- data ready
    err: out std_logic; -- error, input out of range
    ovrflw: out std_logic; -- multipliers or adders overflowed
    func_of_angle: out signed(DO_SIZ-1 downto 0)
);
end SineUsingTaylor;

architecture RTL of SineUsingTaylor is

--   declared constants
type ca_type is array (natural range <>) of signed(M_SIZ-1 downto 0);
constant COEFFICIENTS: ca_type (0 to 4) :=
(
    RealToQmn(1.0/6.0,M_INT_SIZ,M_FRAC_SIZ,TZERO),
    RealToQmn(1.0/120.0,M_INT_SIZ,M_FRAC_SIZ,TZERO),
    RealToQmn(1.0/5040.0,M_INT_SIZ,M_FRAC_SIZ,TZERO),
    RealToQmn(1.0/362880.0,M_INT_SIZ,M_FRAC_SIZ,TZERO),
    RealToQmn(0.0,M_INT_SIZ,M_FRAC_SIZ,TZERO)
);
--   declared signals
signal angle: signed(DI_SIZ-1 downto 0) := (others=>'0');
signal x: signed(DI_SIZ-1 downto 0) := (others=>'0');
signal dx: signed(M_SIZ-1 downto 0) := (others=>'0');
signal dt: signed(M_SIZ-1 downto 0) := (others=>'0');
signal ds: signed(M_SIZ-1 downto 0) := (others=>'0');

signal dx_cnt: natural range 0 to 8 := 0;
signal dt_cnt: natural range 0 to 4 := 0;
signal ds_cnt: natural range 0 to 4 := 0;

signal mult_err1: std_logic := '0';
signal mult_err2: std_logic := '0';

--   master state thread machine enumeration list
type mstr_sm_def is
(
    RESET,START_CNVRT,CHECK_BOUNDS,CREATE_X,
    COMPUTE,COMPUTE2,DONOT_CONVERT,CONVERT_SIGN,
    HANDLE_OVRFLW
);
signal mstr_state, mstr_ret_state: mstr_sm_def := RESET;
-- thread1 state machine enumeration list
type thr1_sm_def is
```

```
(
    THR1_RESET,THR1_START,THR1_COMP_DX,THR1_COMP_DX2,
    THR1_COMP_DX3
);
signal thr1_state: thr1_sm_def := THR1_RESET;
-- thread2 state machine enumeration list
type thr2_sm_def is
(
    THR2_RESET,THR2_START,THR2_COMP_DT,THR2_COMP_DT2,
    THR2_COMP_DT3
);
signal thr2_state: thr2_sm_def := THR2_RESET;
-- thread3 state machine enumeration list
type thr3_sm_def is
(
    THR3_RESET,THR3_START,THR3_COMP_DS,THR3_COMP_DS2
);
signal thr3_state: thr3_sm_def := THR3_RESET;
begin
-------------------------------------------------------------
--   Sine Function state machine (using Radians)
-------------------------------------------------------------
process(rst,clk)
begin
    if(rst='1') then
        -- outputs
        rdy <= '0';err <= '0';ovrflw <= '0';
        mult_err1 <= '0';mult_err2 <= '0';
        func_of_angle <= (others=>'0');
        -- local data objects
        angle <= (others=>'0');
        x <= (others=>'0');
        dx <= (others=>'0');
        dt <= (others=>'0');
        ds <= (others=>'0');
        dx_cnt <= 0;
        dt_cnt <= 0;
        ds_cnt <= 0;
        -- external multipliers
        start1 <= '0';
        multiplier1 <= (others=>'0');
        multiplicand1 <= (others=>'0');
        start2 <= '0';
        multiplier2 <= (others=>'0');
```

```vhdl
    multiplicand2 <= (others=>'0');
    -- master states
    mstr_state <= RESET;
    mstr_ret_state <= RESET;
    -- threads 1-3 states
    thr1_state <= THR1_RESET;
    thr2_state <= THR2_RESET;
    thr3_state <= THR3_RESET;

elsif rising_edge(clk) then
    -- one clock signals
    start1 <= '0';
    start2 <= '0';


    --
    --   Master Thread
    --
    case mstr_state is
        -- reset state
        when RESET =>
            mstr_state <= START_CNVRT;
        -- wait for signal to start conversion
        when START_CNVRT =>
            if(cnvrt = '1') then
                rdy <= '0';err <= '0';ovrflw <= '0';
                angle <= theta;
                mstr_state <= CHECK_BOUNDS;
            end if;
        -- directly assign boundaries of pi, otherwise
        -- compute using taylor polynomials.
        when CHECK_BOUNDS =>
            if(angle > \2pi\ or angle < 0) then
                rdy <= '1';err <= '1';
                mstr_state <= START_CNVRT;
            elsif(angle = \3pi/2\) then
                func_of_angle <= MINUS1;
                rdy <= '1';
                mstr_state <= START_CNVRT;
            elsif(angle = \pi\) then
                func_of_angle <= (others=>'0');
                rdy <= '1';
                mstr_state <= START_CNVRT;
            elsif(angle = \pi/2\) then
                func_of_angle <= PLUS1;
```

```
            rdy <= '1';
            mstr_state <= START_CNVRT;
        elsif(angle = 0) then
            rdy <= '1';
            func_of_angle <= (others=>'0');
            mstr_state <= START_CNVRT;
        else
            mstr_state <= CREATE_X;
        end if;
-- reduce x to 0 to pi/2 for input range of 0 to 2pi.
when CREATE_X =>
    if(angle > \3pi/2\) then
        x <= \pi/2\ - (angle - \3pi/2\);
        mstr_ret_state <= CONVERT_SIGN;
    elsif(angle > \pi\) then
        x <= angle - \pi\;
        mstr_ret_state <= CONVERT_SIGN;
    elsif(angle > \pi/2\) then
        x <= \pi/2\ - (angle - \pi/2\);
        mstr_ret_state <= DONOT_CONVERT;
    else
        x <= angle;
        mstr_ret_state <= DONOT_CONVERT;
    end if;
    mstr_state <= COMPUTE;
-- initiate other threads to compute function
when COMPUTE =>
    mstr_state <= COMPUTE2;
when COMPUTE2 =>
    if(mult_err1 = '1' or mult_err2 = '1') then
        mstr_state <= HANDLE_OVRFLW;
    elsif(dx_cnt = 8 and dt_cnt = 4 and ds_cnt = 4) then
        mstr_state <= mstr_ret_state;
    end if;
when DONOT_CONVERT =>
    func_of_angle <= resize(ds,func_of_angle'length);
    rdy <= '1';
    mstr_state <= START_CNVRT;
when CONVERT_SIGN =>
    func_of_angle <= resize(((not ds) + 1),func_of_angle'length);
    rdy <= '1';
    mstr_state <= START_CNVRT;
when HANDLE_OVRFLW =>
    -- wait for other threads to complete
```

```
            if(thr1_state = THR1_START and
                thr2_state = THR2_START and thr3_state = THR3_START) then
                rdy <= '1';
                ovrflw <= '1';
                mstr_state <= START_CNVRT;
            end if;
        when others =>
            mstr_state <= RESET;
end case;
--
--  Thread 1 (computing x to its nth power)
--
case thr1_state is
    when THR1_RESET =>
        thr1_state <= THR1_START;
    -- start sequence of advancing delta x
    when THR1_START =>
        if(mstr_state = COMPUTE) then
            dx_cnt <= 0;
            dx <= resize(x,dx'length);
            thr1_state <= THR1_COMP_DX;
        end if;
    -- raise x to the next power
    when THR1_COMP_DX =>
        start1 <= '1';
        multiplier1 <= dx;
        multiplicand1 <= resize(x,multiplicand1'length);
        thr1_state <= THR1_COMP_DX2;
    when THR1_COMP_DX2 =>
        -- check for multiplier to complete
        if(cmplt1 = '1') then
            -- abort if overflow
            if(ovrflw1 = '1') then
                thr1_state <= THR1_START;
            -- store results and update counters
            else
                dx <= product1;
                dx_cnt <= dx_cnt + 1;
                thr1_state <= THR1_COMP_DX3;
            end if;
        end if;
    -- check handshake
    when THR1_COMP_DX3 =>
        if(mult_err2 = '1') then
```

```vhdl
                    -- abort if thread2 error
                    thr1_state <= THR1_START;
                else
                    case dx_cnt is
                        -- advance power of x
                        when 1 | 3 | 5 | 7 =>
                            thr1_state <= THR1_COMP_DX;
                        -- wait for next term multiplication to start,
                        -- include dependencies, then advance power of x.
                        when 2 =>
                            if(start2 = '1' and dt_cnt = 0) then
                                thr1_state <= THR1_COMP_DX;
                            end if;
                        when 4 =>
                            if(start2 = '1' and dt_cnt = 1) then
                                thr1_state <= THR1_COMP_DX;
                            end if;
                        when 6 =>
                            if(start2 = '1' and dt_cnt = 2) then
                                thr1_state <= THR1_COMP_DX;
                            end if;
                        when others =>
                            -- thread complete
                            thr1_state <= THR1_START;
                    end case;
                end if;
        when others =>
            .thr1_state <= THR1_RESET;
end case;
--
--   Thread 2 (computing individual terms)
--
case thr2_state is
    when THR2_RESET =>
        thr2_state <= THR2_START;
    -- start sequence of advancing delta t
    when THR2_START =>
        if(mstr_state = COMPUTE) then
            dt_cnt <= 0;
            dt <= (others=>'0');
            thr2_state <= THR2_COMP_DT;
        end if;
    -- check handshake then create next term
    when THR2_COMP_DT =>
```

```
if(mult_err1 = '1') then
    -- abort if error in thread 1
    thr2_state <= THR2_START;
else
    case dx_cnt is
        -- wait
        when 0 | 1 | 3 | 5 | 7 =>
            null;
        -- advance to next term based on dependencies
        when others =>
            if(dx_cnt = 2 and dt_cnt = 0) or
                (dx_cnt = 4 and dt_cnt = 1) or
                (dx_cnt = 6 and dt_cnt = 2) or
                (dx_cnt = 8 and dt_cnt = 3) then
                    start2 <= '1';
                    multiplier2 <= dx;
                    multiplicand2 <= COEFFICIENTS(dt_cnt);
                    thr2_state <= THR2_COMP_DT2;
            end if;
    end case;
end if;
when THR2_COMP_DT2 =>
    -- check for multiplier to complete
    if(cmplt2 = '1') then
        -- abort if overflow
        if(ovrflw2 = '1') then
            thr2_state <= THR2_START;
        -- store results and update counters
        else
            dt <= product2;
            dt_cnt <= dt_cnt + 1;
            thr2_state <= THR2_COMP_DT3;
        end if;
    end if;
-- check for last term
when THR2_COMP_DT3 =>
    if(dt_cnt < 4) then
        thr2_state <= THR2_COMP_DT;
    else
        thr2_state <= THR2_START;
    end if;
when others =>
    thr2_state <= THR2_RESET;
end case;
```

```
--
--   Thread 3 (adding and subtracting terms)
--
case thr3_state is
    when THR3_RESET =>
        thr3_state <= THR3_START;
    -- start sequence of advancing delta s
    when THR3_START =>
        if(mstr_state = COMPUTE) then
            ds_cnt <= 0;
            ds <= resize(x,ds'length);
            thr3_state <= THR3_COMP_DS;
        end if;
    -- wait for each term to be computed
    when THR3_COMP_DS =>
        if(mult_err1 = '1' or mult_err2 = '1') then
            -- abort if error
            thr3_state <= THR3_START;
        elsif(dt_cnt = 1 and ds_cnt = 0) then
            ds <= ds - dt;
            ds_cnt <= ds_cnt + 1;
            thr3_state <= THR3_COMP_DS2;
        elsif(dt_cnt = 2 and ds_cnt = 1) then
            ds <= ds + dt;
            ds_cnt <= ds_cnt + 1;
            thr3_state <= THR3_COMP_DS2;
        elsif(dt_cnt = 3 and ds_cnt = 2) then
            ds <= ds - dt;
            ds_cnt <= ds_cnt + 1;
            thr3_state <= THR3_COMP_DS2;
        elsif(dt_cnt = 4 and ds_cnt = 3) then
            ds <= ds + dt;
            ds_cnt <= ds_cnt + 1;
            thr3_state <= THR3_COMP_DS2;
        end if;
    -- advance count for delta sum
    when THR3_COMP_DS2 =>
        if(ds_cnt < 4) then
            thr3_state <= THR3_COMP_DS;
        else
            thr3_state <= THR3_START;
        end if;
    when others =>
        thr3_state <= THR3_RESET;
```

```
        end case;
        --
        --   Multiplier overflow error detection
        --
        -- multiplier 1
        if(start1 = '1' or mstr_state = START_CNVRT) then
            mult_err1 <= '0';
        elsif(cmplt1 = '1' and ovrflw1 ='1') then
            mult_err1 <= '1';
        end if;
        -- multiplier 2
        if(start2 = '1' or mstr_state = START_CNVRT) then
            mult_err2 <= '0';
        elsif(cmplt2 = '1' and ovrflw2 ='1') then
            mult_err2 <= '1';
        end if;

    end if;
end process;

end RTL;
```

4.1.4 Cosine Function (Taylor-Series)

Taylor-Series notation for cosine in both sigma and expanded form are provided in Figure 8, followed by its graphical representation. The solid line was created by an accurate software algorithm. The dashed line represents the simulated Taylor-Series expression of the 8th degree over π radians.

$$\sum_{n=0}^{4} (-1)^n \frac{x^{2n}}{(2n)!}$$

$$\cos x = 1 - \frac{x^2}{2!} + \frac{x^4}{4!} - \frac{x^6}{6!} + \frac{x^8}{8!}$$

$$(2 \cdot 1)$$
$$(4 \cdot 3 \cdot 2 \cdot 1)$$
$$(6 \cdot 5 \cdot 4 \cdot 3 \cdot 2 \cdot 1)$$
$$(8 \cdot 7 \cdot 6 \cdot 5 \cdot 4 \cdot 3 \cdot 2 \cdot 1)$$

Coefficients from factorials

Figure 8

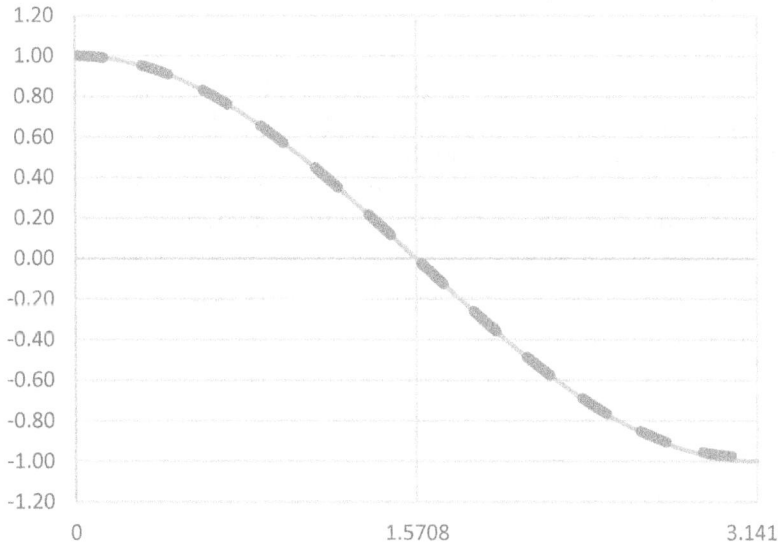

Just as with the sine function, divergence begins to occur slightly before the π radians boundary of 3.1416, lower right-hand corner of the graph. This means that an expansion of the 8th degree is only suitable for $\pi/2$ radians, 1.5708, which is one quadrant of range. This is what we need and the state machine will compute the other quadrants.

Note: *An expansion of 6th degree would be inaccurate when approaching the $\pi/2$ boundary, making it unsuitable for one quadrant.*

Commonality

The cosine design is identical to the sin x design in almost every way, except the value that the data objects are initialized to and the coefficients used. The first point changes the progression of the power of x to x^2, x^4, x^6, and x^8, instead of x^3, x^5, x^7, and x^9 used by the sine function.

Note: *Data objects Δx, Δt, and Δs are initialized to 1, 0, and 1, instead of x, 0, and x.*

Figure 9 represents the thread progression of the Taylor-Series cosine function.

Recap on state machine design for Cosine function (Taylor-Series)

- Design support Q3.32 for a range of 0 to 2π radians input, and Q1.32 for an output ranging from -1.0 to +1.0
- An *err* status bit indicates the input has exceeded 2π radians, an *ovrflw* bit indicates an overflow occurred in one of the external multipliers and the result is invalid.

Taylor-Series Cosine Function Progression

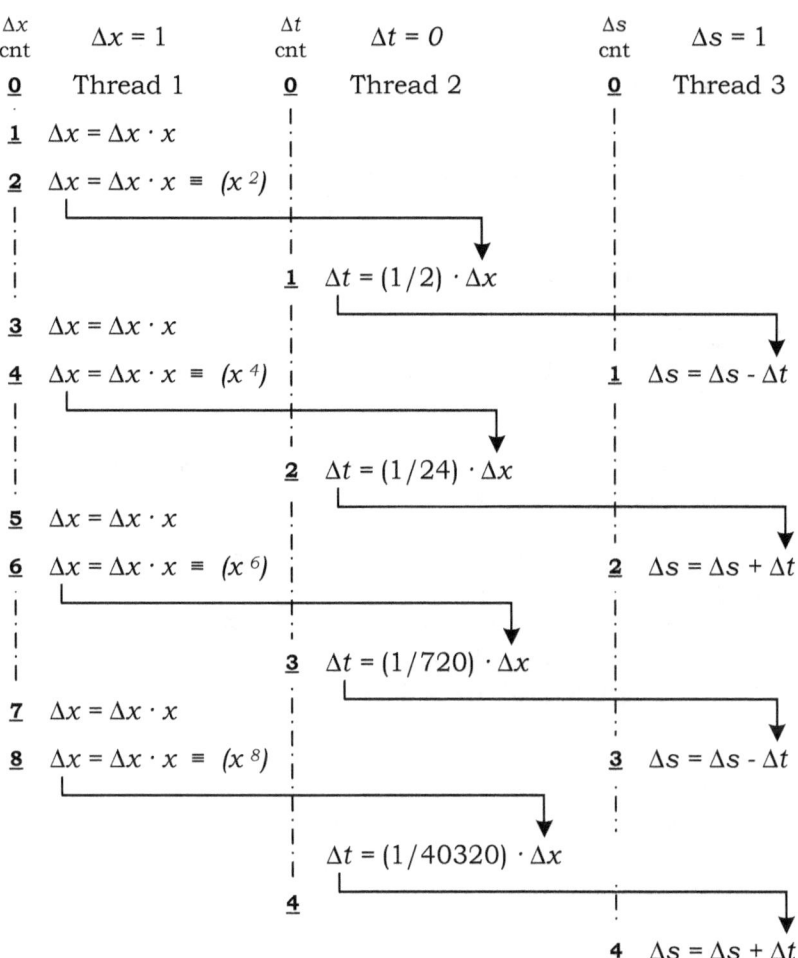

Figure 9

4.1.4.1   Cosine Function (source code)

Note: *Only differences are shown between the Taylor-Series sine function and the cosine function because the source code is virtually identical.*

```
---------------------------------------------------------------
-- CosineUsingTaylor.vhd (radians as input)
---------------------------------------------------------------
.
entity CosineUsingTaylor is
.
end CosineUsingTaylor;
architecture RTL of CosineUsingTaylor is

--  declared constants
type ca_type is array (natural range <>) of signed(M_SIZ-1 downto 0);
constant COEFFICIENTS: ca_type (0 to 4) :=
(
    RealToQmn(1.0/2.0,M_INT_SIZ,M_FRAC_SIZ,TZERO),
    RealToQmn(1.0/24.0,M_INT_SIZ,M_FRAC_SIZ,TZERO),
    RealToQmn(1.0/720.0,M_INT_SIZ,M_FRAC_SIZ,TZERO),
    RealToQmn(1.0/40320.0,M_INT_SIZ,M_FRAC_SIZ,TZERO),
    RealToQmn(0.0,M_INT_SIZ,M_FRAC_SIZ,TZERO)
);
.
.------------------------------------------------
-- Cosine Function state machine (using Radians)
------------------------------------------------
process(rst,clk)
begin

        --
        -- Master Thread
        --
        .
        when CHECK_BOUNDS =>
            if(angle > \2pi\ or angle < 0) then
                rdy <= '1';err <= '1';
                mstr_state <= START_CNVRT;
            elsif(angle = \3pi/2\) then
                func_of_angle <= (others=>'0');
                rdy <= '1';
                mstr_state <= START_CNVRT;
```

```
        elsif(angle = \pi\) then
            func_of_angle <= MINUS1;
            rdy <= '1';
            mstr_state <= START_CNVRT;
        elsif(angle = \pi/2\) then
            func_of_angle <= (others=>'0');
            rdy <= '1';
            mstr_state <= START_CNVRT;
        elsif(angle = 0) then
            rdy <= '1';
            func_of_angle <= PLUS1;
            mstr_state <= START_CNVRT;
        else
            mstr_state <= CREATE_X;
        end if;
    -- reduce x to 0 to pi/2 for input range of 0 to 2pi.
    when CREATE_X =>
        if(angle > \3pi/2\) then
            x <= \pi/2\ - (angle - \3pi/2\);
            mstr_ret_state <= DONOT_CONVERT;
        elsif(angle > \pi\) then
            x <= angle - \pi\;
            mstr_ret_state <= CONVERT_SIGN;
        elsif(angle > \pi/2\) then
            x <= \pi/2\ - (angle - \pi/2\);
            mstr_ret_state <= CONVERT_SIGN;
        else
            x <= angle;
            mstr_ret_state <= DONOT_CONVERT;
        end if;
        mstr_state <= COMPUTE;


--
--  Thread 1 (computing x to its nth power)
--


    -- start sequence of advancing delta x
    when THR1_START =>
        if(mstr_state = COMPUTE) then
            dx_cnt <= 0;
            dx <= RealToQmn(1.0,M_INT_SIZ,M_FRAC_SIZ,TZERO);
            thr1_state <= THR1_COMP_DX;
        end if;
```

```
    --
    --   Thread 2 (computing individual terms)
    --


        -- check handshake then create next term
        when THR2_COMP_DT =>
            if(mult_err1 = '1') then
                -- abort if error in thread 1
                thr2_state <= THR2_START;
            else
                case dx_cnt is
                    -- wait
                    when 0 | 1 | 3 | 5 | 7 =>
                        null;
                    -- advance to next term based on dependencies
                    when others =>
                        if(dx_cnt = 2 and dt_cnt = 0) or
                            (dx_cnt = 4 and dt_cnt = 1) or
                            (dx_cnt = 6 and dt_cnt = 2) or
                            (dx_cnt = 8 and dt_cnt = 3) then
                                start2 <= '1';
                                multiplier2 <= dx;
                                multiplicand2 <= COEFFICIENTS(dt_cnt);
                                thr2_state <= THR2_COMP_DT2;
                        end if;
                end case;
            end if;


    --
    --   Thread 3 (adding and subtracting terms)
    --


        -- start sequence of advancing delta s
        when THR3_START =>
            if(mstr_state = COMPUTE) then
                ds_cnt <= 0;
                ds <= RealToQmn(1.0,M_INT_SIZ,M_FRAC_SIZ,TZERO);

                thr3_state <= THR3_COMP_DS;
            end if;

end if;
end process;
end RTL;
```

4.1.5 Tangent Function (Taylor Series)

There is a Taylor-Series formula for the tangent function, but it is
deemed impractical for hardware implementation. Even with an
order of the 15th degree (eight terms), divergences begins within +/-
2. A better solution by far is to combine sine and cosine.

$$\tan x = \frac{\sin x}{\cos x}$$

At first glance this approach may appear to have excessive
overhead and a lengthy execution time. Fortunately, we can take
advantage of the similarities between sine and cosine, namely the
power of x.

Cosine uses x^2, x^4, x^6, and x^8 within its polynomial terms; whereas
sine uses x^3, x^5, x^7, and x^9. The alternating nature of the powers of
x between cosine and sine are obvious. For example, while
computing the cosine term using x^2, x^3 can be computed for the
next sine term. While computing the sine term using x^3, x^4 can be
computed for the next cosine term. The portion of the tangent
execution time involved in computing both sin x and cos x, is
approximately the same as computing either separately, as done in
the Taylor-Series sine and cosine implementations.

   Note: *The actual implemention uses holding registers, which
   further extends overlap.*

Additional time, however, involves the last step when sin x is
divided by cos x, and the addition of a third compute state in Figure
5. Even so, there are significant advantages:

- Multipliers used throughout the design only require the
  integer/fractional format to support sine or cosine, in
  particular a small integer portion.
- The divider module is independent and its format can be
  scaled to support larger results, directly related to the
  integer format of the divider, without effecting other internal
  data paths. Default is +/-63.9999999998, but can easily be
  larger.

Threads

The master thread for tangent has one extra state and is interfaced to an external divider, otherwise it is the same as those used in sine and cosine functions. Because the computation of sine and cosine is done concurrently, the design is made up of five parts or five threads.

- Thread 1 – advances the value of x to the next nth power, maintained in Δx (delta x). Requires an external multiplier.

- Sine Thread 2 – computes each polynomial term for sine at each stage of the algorithm, sine Δt (delta term). Requires an external multiplier.

- Sine Thread 3 – computes the collective sum as each sine polynomial term is retired, sine Δs (delta sum). Uses a local adder.

- Cosine Thread 2 – computes each polynomial term for cosine at each stage of the algorithm, cosine Δt (delta term). Requires an external multiplier.

- Cosine Thread 3 - computes the collective sum as each cosine polynomial term is retired, cosine Δs (delta sum). Uses a local adder.

As with the sine and cosine designs, each thread is independent, handshaking between threads regulates the flow of data and progress of each thread.

Handshake

The paradigm throughout section 4.1 employs the use of counters and control signals to regulate when data objects are advanced along with their associated counters. Thread 1, which advances the power of x in Δx, with both the sine and cosine threads, is the most complicated.

Note: *in addition to* Δx, *are* sin_Δx, sin_Δt, *and* sin_Δs *for the sine portion, and* cos_Δx, cos_Δt, *and* cos_Δs *for the cosine.*

Figure 10 aptly illustrates the coordination of the power of x with the threads supporting sine and cosine. As Δx advances, so does its associated counter. Depending on its count, the holding register set supporting cosine -- or -- sine is updated with the current Δx and $\Delta x\_$cnt. When the corresponding thread, cosine thread 2 or sine thread 2 sees an updated count the updated Δx is passed to its multiplier to generate the next polynomial term. The same control signal that starts the multiplication (*taken*) clears the holding register count, cos_$\Delta x\_$cnt or sin_$\Delta x\_$cnt.

While thread 1 has already started computing the next power of x for the other thread, clearing of the holding register counter allows thread 1 to post the next value when its ready.

Truth table for computing Thread 1, computing the power of x.

Dependencies			Action/Operation
Δx cnt	cos Δx cnt	sin Δx cnt	
0	X	X	$\Delta x = \Delta x \cdot x$ Increment Δx cnt
1,3,5,7	0	X	$\cos \Delta x = \Delta x$ $\cos \Delta x$ cnt $= \Delta x$ cnt, then $\Delta x = \Delta x \cdot x$ Increment Δx cnt
2,4,6	X	0	$\sin \Delta x = \Delta x$ $\sin \Delta x$ cnt $= \Delta x$ cnt, then $\Delta x = \Delta x \cdot x$ Increment Δx cnt
8	X	0	$\sin \Delta x = \Delta x$ $\sin \Delta x$ cnt $= \Delta x$ cnt

Note: Δx cnt is advanced when Δx is updated with a new value.

Figures 11 and 12 show the progression of cosine threads 2 and 3, as well as sine threads 2 and 3. The tables on the next page show their dependencies. Once all counts have been reached, the master thread completes the operation by dividing sin x by cos x, then assigns the proper sign and sets *rdy*.

Truth table for computing Thread 2, sine polynomial terms.

Dependencies		Action/Operation
sin Δx cnt	sin Δt cnt	
2	0	sin Δt = sine coefficient • sin Δx, then increment sin Δt cnt
4	1	
6	2	
8	3	

Note: *coefficients are supplied by an fixed array, indexed by the current* sin Δt *count.*

Truth table for controlling Thread 3, summing sine terms.

Dependencies		Action/Operation
sin Δt cnt	sin Δs cnt	
1	0	sin Δs = sin Δs - sin Δt, then increment sin Δs cnt
2	1	sin Δs = sin Δs + sin Δt, then increment sin Δs cnt
3	2	sin Δs = sin Δs - sin Δt, then increment sin Δs cnt
4	3	sin Δs = sin Δs + sin Δt, then increment sin Δs cnt

Truth table for computing Thread 2, cosine polynomial terms.

Dependencies		Action/Operation
cos Δx cnt	cos Δt cnt	
1	0	cos Δt = cosine coefficient • cos Δx, then increment cos Δt cnt
3	1	
5	2	
7	3	

Note: *coefficients are supplied by a fixed array, indexed by the current* cos Δt *count.*

Truth table for controlling Thread 3, summing cosine terms.

Dependencies		Action/Operation
cos Δt cnt	cos Δs cnt	
1	0	cos Δs = cos Δs - cos Δt, then increment cos Δs cnt
2	1	cos Δs = cos Δs + cos Δt, then increment cos Δs cnt
3	2	cos Δs = cos Δs - cos Δt, then increment cos Δs cnt
4	3	cos Δs = cos Δs + cos Δt, then increment cos Δs cnt

Recap of state machine design for the Tangent function (Taylor-Series)

- Design support Q3.32 for a range of 0 to 2π radians input.
- Divider defaults Q6.32 resulting an output from +/- 63.9999999998. When the external divider signals an overflow from the two operands the output is set to its saturated value and the *ovr* bit is set true.
- The *err* status bit indicating the input has exceeded 2π radians, an *ovrflw* bit indicates an overflow occurred in one of the external multipliers or divider, *udrflw* indicate an underflow occurred in the divider. Either means the output data is invalid.
- Bit *udf* indicates undefined region of tangent boundary.

Taylor-Series Tangent Thread Progression 1

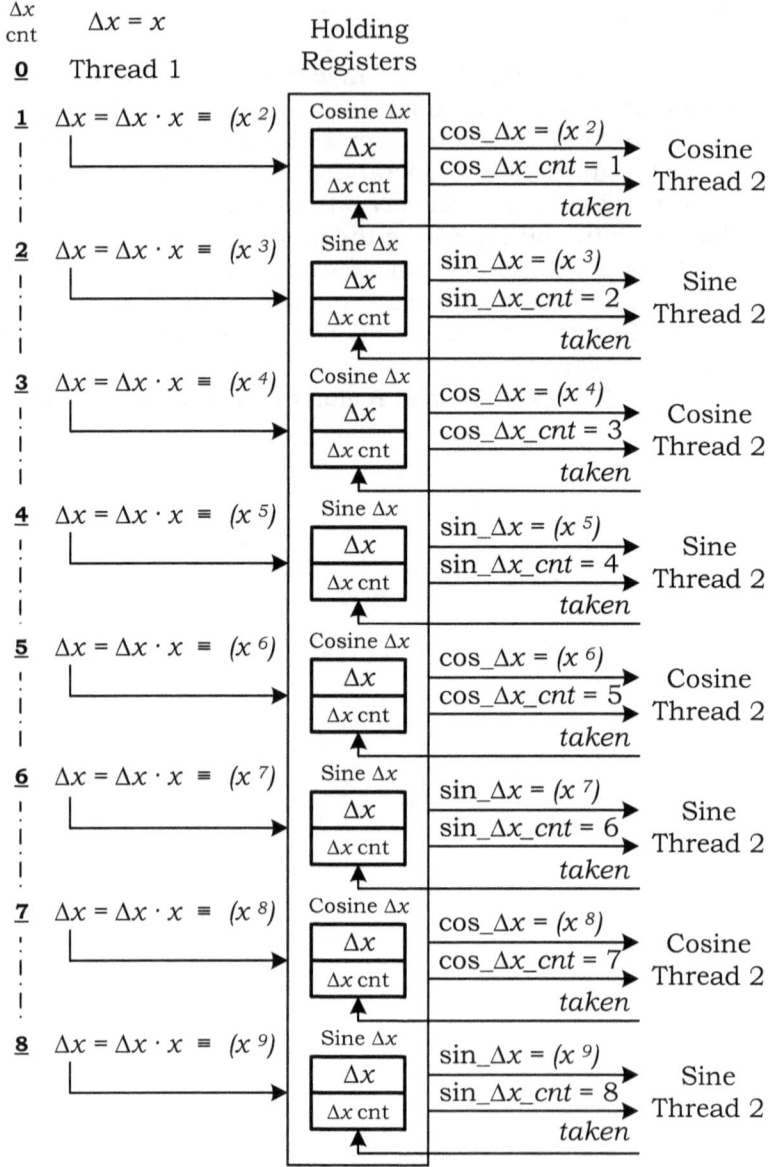

Figure 10

Taylor-Series Tangent Function Progression 2

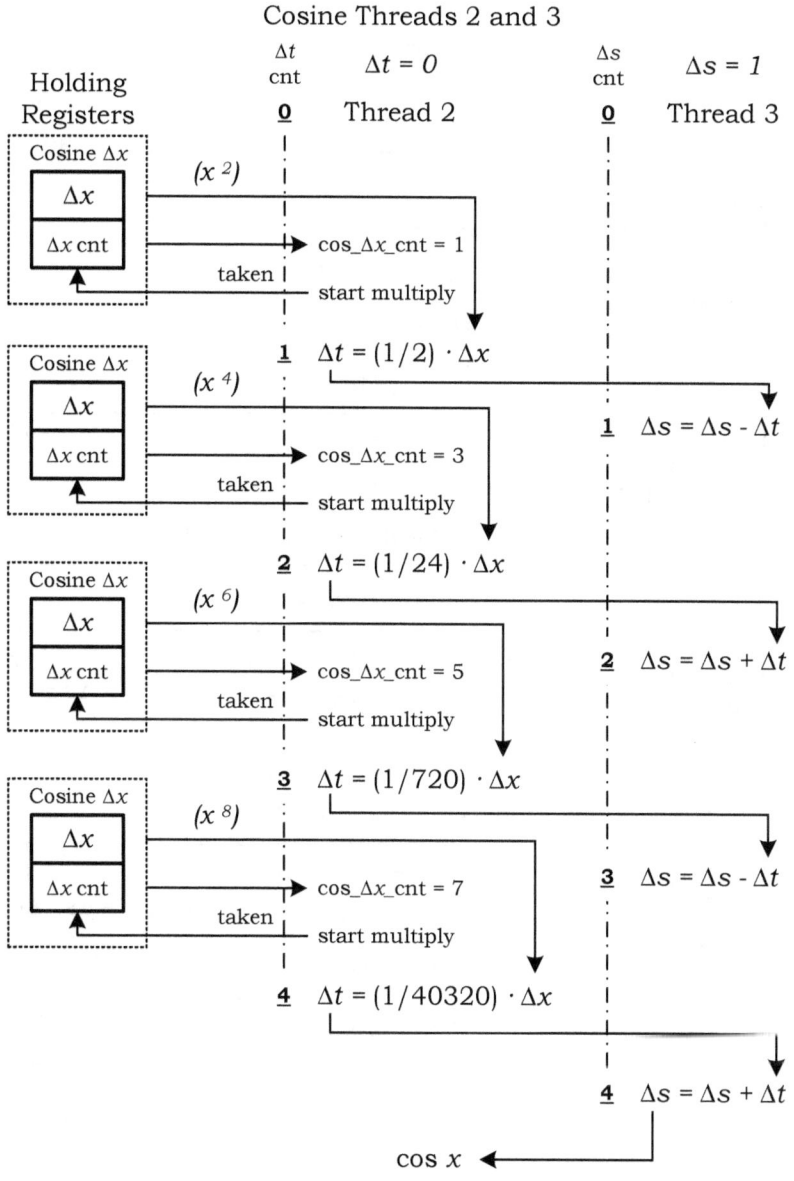

Cosine Threads 2 and 3

Figure 11

Taylor-Series Tangent Function Progression 3

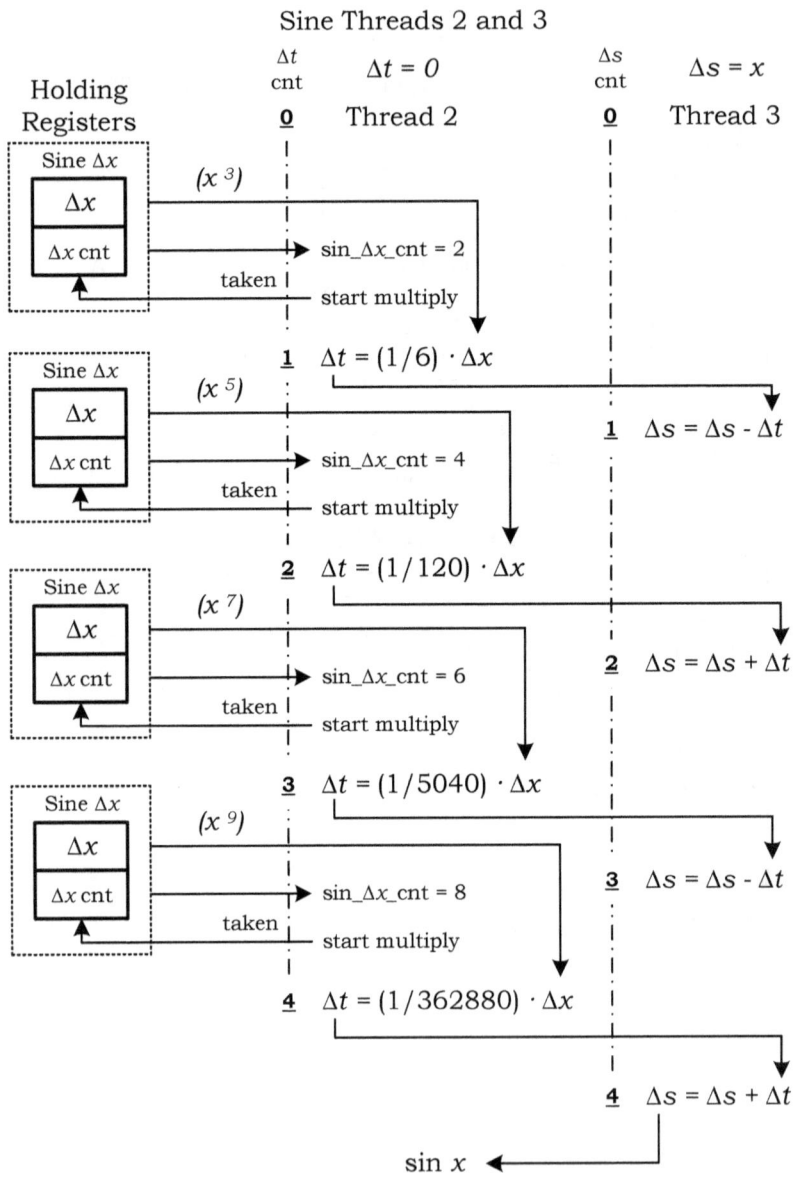

Figure 12

4.1.5.1   Tangent Function (source code)

```
-----------------------------------------------------------------------------
--   TangentUsingTaylor.vhd (radians as input)
-----------------------------------------------------------------------------
library IEEE;
use IEEE.std_logic_1164.all;
use IEEE.numeric_std.all;
use IEEE.math_real.all;
-- user packages
use work.ConversionPackageV2.all;
use work.TaylorRealTimePackage.all;

entity TangentUsingTaylor is
--   Input and output fixed-point formats, as well as internal data paths,
--   are defined in TaylorRealTimePackage.vhd
port
(
    clk,rst: in std_logic; -- system clock and reset (synchronous)
    -- inputs
    cnvrt: in std_logic; -- initiate tangent function
    theta: in signed(DI_SIZ-1 downto 0); -- theta in radians
    -- interface to external multiplier 1
    start1: inout std_logic; -- start multiplication
    multiplier1: out signed(M_SIZ-1 downto 0);
    multiplicand1: out signed(M_SIZ-1 downto 0);
    cmplt1: in std_logic; -- multiplication complete
    ovrflw1: in std_logic; -- overflow error
    product1: in signed(M_SIZ-1 downto 0);
    -- interface to external multiplier 2
    start2: inout std_logic; -- start multiplication
    multiplier2: out signed(M_SIZ-1 downto 0);
    multiplicand2: out signed(M_SIZ-1 downto 0);
    cmplt2: in std_logic; -- multiplication complete
    ovrflw2: in std_logic; -- overflow error
    product2: in signed(M_SIZ-1 downto 0);
    -- interface to external multiplier 3
    start3: inout std_logic; -- start multiplication
    multiplier3: out signed(M_SIZ-1 downto 0);
    multiplicand3: out signed(M_SIZ-1 downto 0);
    cmplt3: in std_logic; -- multiplication complete
    ovrflw3: in std_logic; -- overflow error
    product3: in signed(M_SIZ-1 downto 0);
    -- interface to external divider
```

153

```vhdl
        start4: out std_logic; -- start division
        divisor4: out signed(D_SIZ-1 downto 0);
        dividend4: out signed(D_SIZ-1 downto 0);
        cmplt4: in std_logic; -- division complete
        ovrflw4: in std_logic; -- overflow error
        udrflow4: in std_logic; -- underflow error
        quotient4: in signed(D_SIZ-1 downto 0);
        -- outputs
        rdy: out std_logic; -- data ready
        err: out std_logic; -- error, input out of range
        ovr: out std_logic; -- output at its maximum range, saturated
        udf: out std_logic; -- undefined at pi/2 or 3pi/2 radians boundaries
        ovrflw: out std_logic; -- multipliers or adders overflowed
        udrflow: out std_logic; -- division underflow
        func_of_angle: out signed(D_SIZ-1 downto 0) -- same as divider output
);
end TangentUsingTaylor;

architecture RTL of TangentUsingTaylor is

--   declared constants
type ca_type is array (natural range <>) of signed(M_SIZ-1 downto 0);
constant SIN_COEFFICIENTS: ca_type (0 to 4) :=
(
    RealToQmn(1.0/6.0,M_INT_SIZ,M_FRAC_SIZ,TZERO),
    RealToQmn(1.0/120.0,M_INT_SIZ,M_FRAC_SIZ,TZERO),
    RealToQmn(1.0/5040.0,M_INT_SIZ,M_FRAC_SIZ,TZERO),
    RealToQmn(1.0/362880.0,M_INT_SIZ,M_FRAC_SIZ,TZERO),
    RealToQmn(0.0,M_INT_SIZ,M_FRAC_SIZ,TZERO)
);
constant COS_COEFFICIENTS: ca_type (0 to 4) :=
(
    RealToQmn(1.0/2.0,M_INT_SIZ,M_FRAC_SIZ,TZERO),
    RealToQmn(1.0/24.0,M_INT_SIZ,M_FRAC_SIZ,TZERO),
    RealToQmn(1.0/720.0,M_INT_SIZ,M_FRAC_SIZ,TZERO),
    RealToQmn(1.0/40320.0,M_INT_SIZ,M_FRAC_SIZ,TZERO),
    RealToQmn(0.0,M_INT_SIZ,M_FRAC_SIZ,TZERO)
);
--   declared signals
signal angle: signed(DI_SIZ-1 downto 0) := (others=>'0');
signal x: signed(DI_SIZ-1 downto 0) := (others=>'0');
signal dx: signed(M_SIZ-1 downto 0) := (others=>'0');
signal dx_cnt: natural range 0 to 8 := 0;
signal result: signed(D_SIZ-1 downto 0) := (others=>'0');
```

```
-- for sine computation
signal sin_dx: signed(M_SIZ-1 downto 0) := (others=>'0');
signal sin_dt: signed(M_SIZ-1 downto 0) := (others=>'0');
signal sin_ds: signed(M_SIZ-1 downto 0) := (others=>'0');
signal sin_dx_cnt: natural range 0 to 8 := 0;
signal sin_dt_cnt: natural range 0 to 4 := 0;
signal sin_ds_cnt: natural range 0 to 4 := 0;
-- for cosine computation
signal cos_dx: signed(M_SIZ-1 downto 0) := (others=>'0');
signal cos_dt: signed(M_SIZ-1 downto 0) := (others=>'0');
signal cos_ds: signed(M_SIZ-1 downto 0) := (others=>'0');
signal cos_dx_cnt: natural range 0 to 8 := 0;
signal cos_dt_cnt: natural range 0 to 4 := 0;
signal cos_ds_cnt: natural range 0 to 4 := 0;

signal mult_err1: std_logic := '0';
signal mult_err2: std_logic := '0';
signal mult_err3: std_logic := '0';

--   master state thread machine enumeration list
type mstr_sm_def is
(
    RESET,START_CNVRT,CHECK_BOUNDS,CREATE_X,COMPUTE,
    COMPUTE2,  COMPUTE3,DONOT_CONVERT,CONVERT_SIGN,
    HANDLE_OVRFLW
);
signal mstr_state, mstr_ret_state: mstr_sm_def := RESET;
-- thread1 state machine enumeration list
type thr1_sm_def is
(
    THR1_RESET,THR1_START,THR1_COMP_DX,THR1_COMP_DX2,
    THR1_COMP_DX3
);
signal thr1_state: thr1_sm_def := THR1_RESET;
-- sine thread2 state machine enumeration list
type sin_thr2_sm_def is
(
    SIN_THR2_RESET,SIN_THR2_START,SIN_THR2_COMP_DT,
    SIN_THR2_COMP_DT2,SIN_THR2_COMP_DT3
);
signal sin_thr2_state: sin_thr2_sm_def := SIN_THR2_RESET;
-- sine thread3 state machine enumeration list
type sin_thr3_sm_def is
(
```

```
    SIN_THR3_RESET,SIN_THR3_START,SIN_THR3_COMP_DS,
    SIN_THR3_COMP_DS2
);
signal sin_thr3_state: sin_thr3_sm_def := SIN_THR3_RESET;
-- cosine thread2 state machine enumeration list
type cos_thr2_sm_def is
(
    COS_THR2_RESET,COS_THR2_START,COS_THR2_COMP_DT,
    COS_THR2_COMP_DT2,COS_THR2_COMP_DT3
);
signal cos_thr2_state: cos_thr2_sm_def := COS_THR2_RESET;
-- cosine thread3 state machine enumeration list
type cos_thr3_sm_def is
(
    COS_THR3_RESET,COS_THR3_START,COS_THR3_COMP_DS,
    COS_THR3_COMP_DS2
);
signal cos_thr3_state: cos_thr3_sm_def := COS_THR3_RESET;

begin
-------------------------------------------------
--   Tangent Function state machine (using Radians)
-------------------------------------------------
process(rst,clk)
begin
    if(rst='1') then
        -- outputs
        rdy <= '0';err <= '0';ovr <= '0';udf <= '0';
        ovrflw <= '0';udrflow <= '0';
        mult_err1 <= '0';mult_err2 <= '0';mult_err3 <= '0';
        func_of_angle <= (others=>'0');
        -- local data objects
        angle <= (others=>'0');
        x <= (others=>'0');
        dx <= (others=>'0');
        dx_cnt <= 0;
        sin_dt <= (others=>'0');
        sin_ds <= (others=>'0');
        sin_dx_cnt <= 0;
        sin_dt_cnt <= 0;
        sin_ds_cnt <= 0;
        cos_dt <= (others=>'0');
        cos_ds <= (others=>'0');
        cos_dx_cnt <= 0;
```

```
    cos_dt_cnt <= 0;
    cos_ds_cnt <= 0;
    -- external multipliers
    start1 <= '0';
    multiplier1 <= (others=>'0');
    multiplicand1 <= (others=>'0');
    start2 <= '0';
    multiplier2 <= (others=>'0');
    multiplicand2 <= (others=>'0');
    start3 <= '0';
    multiplier3 <= (others=>'0');
    multiplicand3 <= (others=>'0');
    -- external divider
    start4 <= '0';
    divisor4 <= (others=>'0');
    dividend4 <= (others=>'0');
    result <= (others=>'0');
    -- master states
    mstr_state <= RESET;
    mstr_ret_state <= RESET;
    -- threads 1-3 states
    thr1_state <= THR1_RESET;
    sin_thr2_state <= SIN_THR2_RESET;
    sin_thr3_state <= SIN_THR3_RESET;
    cos_thr2_state <= COS_THR2_RESET;
    cos_thr3_state <= COS_THR3_RESET;

elsif rising_edge(clk) then
    -- one clock signals
    start1 <= '0';
    start2 <= '0';
    start3 <= '0';
    start4 <= '0';


    --
    --   Master Thread
    --
    case mstr_state is
        -- reset state
        when RESET =>
            mstr_state <= START_CNVRT;
        -- wait for signal to start conversion
        when START_CNVRT =>
            if(cnvrt = '1') then
```

```
            rdy <= '0';err <= '0';ovr <= '0';udf <= '0';
            ovrflw <= '0';udrflow <= '0';
            angle <= theta;
            mstr_state <= CHECK_BOUNDS;
        end if;
-- directly assign boundaries of pi, otherwise
-- compute using taylor polynomials.
when CHECK_BOUNDS =>
    if(angle > \2pi\ or angle < 0) then
            rdy <= '1';err <= '1';
            mstr_state <= START_CNVRT;
    elsif(angle = \3pi/2\) then
            func_of_angle <= (others=>'0'); -- undefined
            rdy <= '1';udf <= '1';
            mstr_state <= START_CNVRT;
    elsif(angle = \pi\) then
            func_of_angle <= (others=>'0'); -- zero
            rdy <= '1';
            mstr_state <= START_CNVRT;
    elsif(angle = \pi/2\) then
            func_of_angle <= (others=>'0'); -- undefined
            rdy <= '1';udf <= '1';
            mstr_state <= START_CNVRT;
    elsif(angle = 0) then
            rdy <= '1';
            func_of_angle <= (others=>'0');  -- zero
            mstr_state <= START_CNVRT;
    else
            mstr_state <= CREATE_X;
    end if;
-- reduce x to 0 to pi/2 for input range of 0 to 2pi.
when CREATE_X =>
    if(angle > \3pi/2\) then
            x <= \pi/2\ - (angle - \3pi/2\);
            mstr_ret_state <= CONVERT_SIGN;
    elsif(angle > \pi\) then
            x <= angle - \pi\;
            mstr_ret_state <= DONOT_CONVERT;
    elsif(angle > \pi/2\) then
            x <= \pi/2\ - (angle - \pi/2\);
            mstr_ret_state <= CONVERT_SIGN;
    else
            x <= angle;
            mstr_ret_state <= DONOT_CONVERT;
```

```
        end if;
        mstr_state <= COMPUTE;
-- initiate other threads to compute sine and cosine functions
when COMPUTE =>
        mstr_state <= COMPUTE2;
when COMPUTE2 =>
        if(mult_err1 = '1' or mult_err2 = '1' or mult_err3 = '1') then
            mstr_state <= HANDLE_OVRFLW;
        -- last step is to divide sin x by cos x
        elsif(dx_cnt = 8 and
            sin_dt_cnt = 4 and sin_ds_cnt = 4 and
            cos_dt_cnt = 4 and cos_ds_cnt = 4) then
            start4 <= '1';
            dividend4 <= sin_ds;
            divisor4 <= cos_ds;
            mstr_state <= COMPUTE3;
        end if;
when COMPUTE3 =>
        -- wait for division to complete
        if(cmplt4 = '1') then
            if(ovrflw4 = '1') then
                -- staturate output
                result <= max_pos_qmn(D_SIZ);
                ovr <= '1';
                mstr_state <= mstr_ret_state;
            elsif(udrflow4 = '1') then
                udrflow <= '1';
                result <= (others=>'0');
                mstr_state <= DONOT_CONVERT;
            else
                -- adjust sign of result
                result <= quotient4;
                mstr_state <= mstr_ret_state;
            end if;
        end if;
when DONOT_CONVERT =>
        func_of_angle <= result;
        rdy <= '1';
        mstr_state <= START_CNVRT;
when CONVERT_SIGN =>
        func_of_angle <= (not result) + 1;
        rdy <= '1';
        mstr_state <= START_CNVRT;
when HANDLE_OVRFLW =>
```

```vhdl
            -- wait for other threads to complete
            if(thr1_state = THR1_START and
                sin_thr2_state = SIN_THR2_START and
                sin_thr3_state = SIN_THR3_START and
                cos_thr2_state = COS_THR2_START and
                cos_thr3_state = COS_THR3_START) then
                rdy <= '1';
                ovrflw <= '1';
                mstr_state <= START_CNVRT;
            end if;
        when others =>
            mstr_state <= RESET;
end case;
--
--   Thread 1 (computing x to its nth power)
--
case thr1_state is
    when THR1_RESET =>
        thr1_state <= THR1_START;
    -- start sequence of advancing delta x
    when THR1_START =>
        if(mstr_state = COMPUTE) then
            dx_cnt <= 0;
            dx <= resize(x,dx'length);
            thr1_state <= THR1_COMP_DX;
        end if;
    -- raise x to the next power
    when THR1_COMP_DX =>
        start1 <= '1';
        multiplier1 <= dx;
        multiplicand1 <= resize(x,multiplicand1'length);
        thr1_state <= THR1_COMP_DX2;
    when THR1_COMP_DX2 =>
        -- check for multiplier to complete
        if(cmplt1 = '1') then
            -- abort if overflow
            if(ovrflw1 = '1') then
                thr1_state <= THR1_START;
            -- store results and update counters
            else
                dx <= product1;
                dx_cnt <= dx_cnt + 1;
                thr1_state <= THR1_COMP_DX3;
            end if;
```

```
            end if;
-- handshake between thread 1 and threads 2 of sine and cosine
    when THR1_COMP_DX3 =>
        if(mult_err2 = '1' or mult_err3 = '1') then
            -- abort if thread2 or 3 error
            thr1_state <= THR1_START;
        else
            case dx_cnt is
                when 1 | 3 | 5 | 7 =>
                    -- make sure previous cos_dx was taken
                    if(cos_dx_cnt = 0) then
                        cos_dx <= dx;
                        cos_dx_cnt <= dx_cnt;
                        thr1_state <= THR1_COMP_DX;
                    end if;
                when 2 | 4 | 6 | 8 =>
                    -- make sure previous sin_dx was taken
                    if(sin_dx_cnt = 0) then
                        sin_dx <= dx;
                        sin_dx_cnt <= dx_cnt;
                        if(dx_cnt = 8) then
                            -- thread complete
                            thr1_state <= THR1_START;
                        else
                            thr1_state <= THR1_COMP_DX;
                        end if;
                    end if;
                when others => null;
            end case;
        end if;
    when others =>
        thr1_state <= THR1_RESET;
end case;
-- resetting Count Registers (part of handshake with thread1
-- THR1_COMP_DX3).
if(start2 = '1') then
    -- multiplication for sine thread 2 started, taken
    sin_dx_cnt <= 0;
end if;
if(start3 = '1') then
    -- multiplication for cosine thread 2 started, taken
    cos_dx_cnt <= 0;
end if;
--
```

```
--  Sine Thread 2 (computing individual terms)
--
case sin_thr2_state is
    when SIN_THR2_RESET =>
        sin_thr2_state <= SIN_THR2_START;
    -- start sequence of advancing delta t
    when SIN_THR2_START =>
        if(mstr_state = COMPUTE) then
            sin_dt_cnt <= 0;
            sin_dt <= (others=>'0');
            sin_thr2_state <= SIN_THR2_COMP_DT;
        end if;
    -- check handshake then create next term
    when SIN_THR2_COMP_DT =>
        if(mult_err1 = '1' or mult_err3 = '1') then
            -- abort if error in thread 1
            sin_thr2_state <= SIN_THR2_START;
        else
                -- advance to next term based on dependencies
                if(sin_dx_cnt = 2 and sin_dt_cnt = 0) or
                    (sin_dx_cnt = 4 and sin_dt_cnt = 1) or
                    (sin_dx_cnt = 6 and sin_dt_cnt = 2) or
                    (sin_dx_cnt = 8 and sin_dt_cnt = 3) then
                        start2 <= '1';
                        multiplier2 <= sin_dx;
                        multiplicand2 <= SIN_COEFFICIENTS(sin_dt_cnt);
                        sin_thr2_state <= SIN_THR2_COMP_DT2;
                end if;
        end if;
    when SIN_THR2_COMP_DT2 =>
        -- check for multiplier to complete
        if(cmplt2 = '1') then
            -- abort if overflow
            if(ovrflw2 = '1') then
                sin_thr2_state <= SIN_THR2_START;
            -- store results and update counters
            else
                sin_dt <= product2;
                sin_dt_cnt <= sin_dt_cnt + 1;
                sin_thr2_state <= SIN_THR2_COMP_DT3;
            end if;
        end if;
    -- check for last term
    when SIN_THR2_COMP_DT3 =>
```

```
        if(sin_dt_cnt < 4) then
            sin_thr2_state <= SIN_THR2_COMP_DT;
        else
            sin_thr2_state <= SIN_THR2_START;
        end if;
    when others =>
        sin_thr2_state <= SIN_THR2_RESET;
end case;
--
--  Sine Thread 3 (adding and subtracting terms)
--
case sin_thr3_state is
    when SIN_THR3_RESET =>
        sin_thr3_state <= SIN_THR3_START;
    -- start sequence of advancing delta s
    when SIN_THR3_START =>
        if(mstr_state = COMPUTE) then
            sin_ds_cnt <= 0;
            sin_ds <= resize(x,sin_ds'length);
            sin_thr3_state <= SIN_THR3_COMP_DS;
        end if;
    -- wait for each term to be computed
    when SIN_THR3_COMP_DS =>
        if(mult_err1 = '1' or mult_err2 = '1' or mult_err3 = '1') then
            -- abort if error
            sin_thr3_state <= SIN_THR3_START;
        elsif(sin_dt_cnt = 1 and sin_ds_cnt = 0) then
            sin_ds <= sin_ds - sin_dt;
            sin_ds_cnt <= sin_ds_cnt + 1;
            sin_thr3_state <= SIN_THR3_COMP_DS2;
        elsif(sin_dt_cnt = 2 and sin_ds_cnt = 1) then
            sin_ds <= sin_ds + sin_dt;
            sin_ds_cnt <= sin_ds_cnt + 1;
            sin_thr3_state <= SIN_THR3_COMP_DS2;
        elsif(sin_dt_cnt = 3 and sin_ds_cnt = 2) then
            sin_ds <= sin_ds - sin_dt;
            sin_ds_cnt <= sin_ds_cnt + 1;
            sin_thr3_state <= SIN_THR3_COMP_DS2;
        elsif(sin_dt_cnt = 4 and sin_ds_cnt = 3) then
            sin_ds <= sin_ds + sin_dt;
            sin_ds_cnt <= sin_ds_cnt + 1;
            sin_thr3_state <= SIN_THR3_COMP_DS2;
        end if;
    -- advance count for delta sum
```

```vhdl
    when SIN_THR3_COMP_DS2 =>
        if(sin_ds_cnt < 4) then
            sin_thr3_state <= SIN_THR3_COMP_DS;
        else
            sin_thr3_state <= SIN_THR3_START;
        end if;
    when others =>
        sin_thr3_state <= SIN_THR3_RESET;
end case;
--
--  Cosine Thread 2 (computing individual terms)
--
case cos_thr2_state is
    when COS_THR2_RESET =>
        cos_thr2_state <= COS_THR2_START;
    -- start sequence of advancing delta t
    when COS_THR2_START =>
        if(mstr_state = COMPUTE) then
            cos_dt_cnt <= 0;
            cos_dt <= (others=>'0');
            cos_thr2_state <= COS_THR2_COMP_DT;
        end if;
    -- check handshake then create next term
    when COS_THR2_COMP_DT =>
        if(mult_err1 = '1' or mult_err2 = '1') then
            -- abort if error in thread 1 or 2
            cos_thr2_state <= COS_THR2_START;
        else
                -- advance to next term based on dependencies
                if(cos_dx_cnt = 1 and cos_dt_cnt = 0) or
                    (cos_dx_cnt = 3 and cos_dt_cnt = 1) or
                    (cos_dx_cnt = 5 and cos_dt_cnt = 2) or
                    (cos_dx_cnt = 7 and cos_dt_cnt = 3) then
                        start3 <= '1';
                        multiplier3 <= cos_dx;
                        multiplicand3 <= COS_COEFFICIENTS(cos_dt_cnt);
                        cos_thr2_state <= COS_THR2_COMP_DT2;
                end if;
        end if;
    when COS_THR2_COMP_DT2 =>
        -- check for multiplier to complete
        if(cmplt3 = '1') then
            -- abort if overflow
            if(ovrflw3 = '1') then
```

```
                cos_thr2_state <= COS_THR2_START;
            -- store results and update counters
            else
                cos_dt <= product3;
                cos_dt_cnt <= cos_dt_cnt + 1;
                cos_thr2_state <= COS_THR2_COMP_DT3;
            end if;
        end if;
    -- check for last term
    when COS_THR2_COMP_DT3 =>
        if(cos_dt_cnt < 4) then
            cos_thr2_state <= COS_THR2_COMP_DT;
        else
            cos_thr2_state <= COS_THR2_START;
        end if;
    when others =>
        cos_thr2_state <= COS_THR2_RESET;
end case;
--
--   Cosine Thread 3 (adding and subtracting terms)
--
case cos_thr3_state is
    when COS_THR3_RESET =>
        cos_thr3_state <= COS_THR3_START;
    -- start sequence of advancing delta s
    when COS_THR3_START =>
        if(mstr_state = COMPUTE) then
            cos_ds_cnt <= 0;
            cos_ds <= RealToQmn(1.0,M_INT_SIZ,M_FRAC_SIZ,TZERO);
            cos_thr3_state <= COS_THR3_COMP_DS;
        end if;
    -- wait for each term to be computed
    when COS_THR3_COMP_DS =>
        if(mult_err1 = '1' or mult_err2 = '1' or mult_err3 = '1') then
            -- abort if error
            cos_thr3_state <= COS_THR3_START;
        elsif(cos_dt_cnt = 1 and cos_ds_cnt = 0) then
            cos_ds <= cos_ds - cos_dt;
            cos_ds_cnt <= cos_ds_cnt + 1;
            cos_thr3_state <= COS_THR3_COMP_DS2;
        elsif(cos_dt_cnt = 2 and cos_ds_cnt = 1) then
            cos_ds <= cos_ds + cos_dt;
            cos_ds_cnt <= cos_ds_cnt + 1;
            cos_thr3_state <= COS_THR3_COMP_DS2;
```

```vhdl
                elsif(cos_dt_cnt = 3 and cos_ds_cnt = 2) then
                    cos_ds <= cos_ds - cos_dt;
                    cos_ds_cnt <= sin_ds_cnt + 1;
                    cos_thr3_state <= COS_THR3_COMP_DS2;
                elsif(cos_dt_cnt = 4 and cos_ds_cnt = 3) then
                    cos_ds <= cos_ds + cos_dt;
                    cos_ds_cnt <= cos_ds_cnt + 1;
                    cos_thr3_state <= COS_THR3_COMP_DS2;
                end if;
            -- advance count for delta sum
            when COS_THR3_COMP_DS2 =>
                if(cos_ds_cnt < 4) then
                    cos_thr3_state <= COS_THR3_COMP_DS;
                else
                    cos_thr3_state <= COS_THR3_START;
                end if;
            when others =>
                cos_thr3_state <= COS_THR3_RESET;
        end case;
        --
        --   Multiplier overflow error detection
        --
        -- multiplier 1
        if(start1 = '1' or mstr_state = START_CNVRT) then
            mult_err1 <= '0';
        elsif(cmplt1 = '1' and ovrflw1 ='1') then
            mult_err1 <= '1';
        end if;
        -- multiplier 2
        if(start2 = '1' or mstr_state = START_CNVRT) then
            mult_err2 <= '0';
        elsif(cmplt2 = '1' and ovrflw2 ='1') then
            mult_err2 <= '1';
        end if;
        -- multiplier 3
        if(start3 = '1' or mstr_state = START_CNVRT) then
            mult_err3 <= '0';
        elsif(cmplt3 = '1' and ovrflw3 ='1') then
            mult_err3 <= '1';
        end if;

    end if;
end process;
end RTL;
```

4.2 Borchardt-Gauss Iteration

Calculation of an arithmetic *mean* involves summing a set of numbers, then dividing that sum by the number of operands in the set. Geometric *mean* is the square root of the product of that number set (factors). If these two *means* are combined and progress by passing the results to the next iteration, their values will converge to the same number. This in essence is a sequence called the Arithmetic-Geometric Mean (AGM).

Borchardt-Gauss mean (BG) is a variant of AGM. The nomenclature for its expression is:

$$BG(a,g)$$

$$a_n+1 = \frac{1}{2}(a_n + g_n) \qquad g_n+1 = \sqrt{(a_n+1)\cdot g_n}$$

Parameter '*a*' represents the arithmetic component and '*g*' the geometric. The algorithm in Figure 13 shows what occurs during each iteration. After successive cycles a_n+1 and g_n+1 will begin to converge, become equal. At which time a final step is executed with the result.

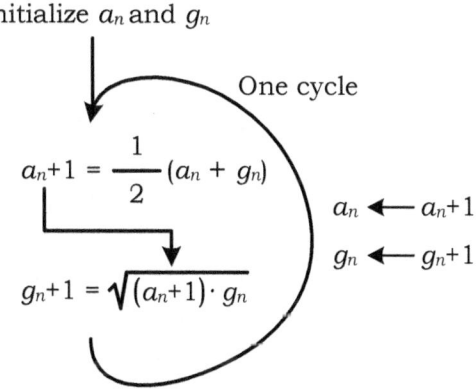

Initialize a_n and g_n

One cycle

$$a_n+1 = \frac{1}{2}(a_n + g_n)$$

$$g_n+1 = \sqrt{(a_n+1)\cdot g_n}$$

$$a_n \leftarrow a_n+1$$

$$g_n \leftarrow g_n+1$$

Exit if a_{n+1} and g_{n+1} are equal

Post Process

Figure 13

This algorithm is identical for computing arcsine, arccosine, and arctangent. The only differences are what 'a' and 'g' are initialized to, what occurs in the final step following convergence, and data path widths that are required for supporting different size operands.

Initial values for each function.

Functions	Initial value $n=0$	
	a_n	G_n
arcsine	$\sqrt{1-x^2}$	1
arccosine	x	1
arctangent	1	$\sqrt{1+x^2}$

Post processing step (division)

Functions	Final Step
arcsine	x/a_n+1
arccosine	$\sqrt{1-x^2}/a_n+1$
arctangent	x/a_n+1

Dissimilar to all other designs in this book, the full range from input to output is computed by this algorithm (the exception being some domain boundaries). All other designs use a single quadrant to compute other quadrants.

4.2.1 Common Features

Sequence

Unlike designs utilizing Taylor-Series, only a single threaded state machine can be used with Borchardt-Gauss. As represented in Figure 13, each step can only proceed once the previous step has completed. In other words, each step has a dependency. There can be no parallel processing.

The good news is that this book series "STATE MACHINES IN VHDL" is intended to expose the reader to state machine designs, and in this section, the use of calls and subroutines. This concept is covered at length in "STATE MACHINES IN VHDL *Composition* Vol. 1", section 2.7 Stacks.

Figure 14 shows the common sequence of all three designs. The first step is make sure the input data is within in the range of the function's domain restrictions, and in some cases, assert a constant value on a domain boundary.

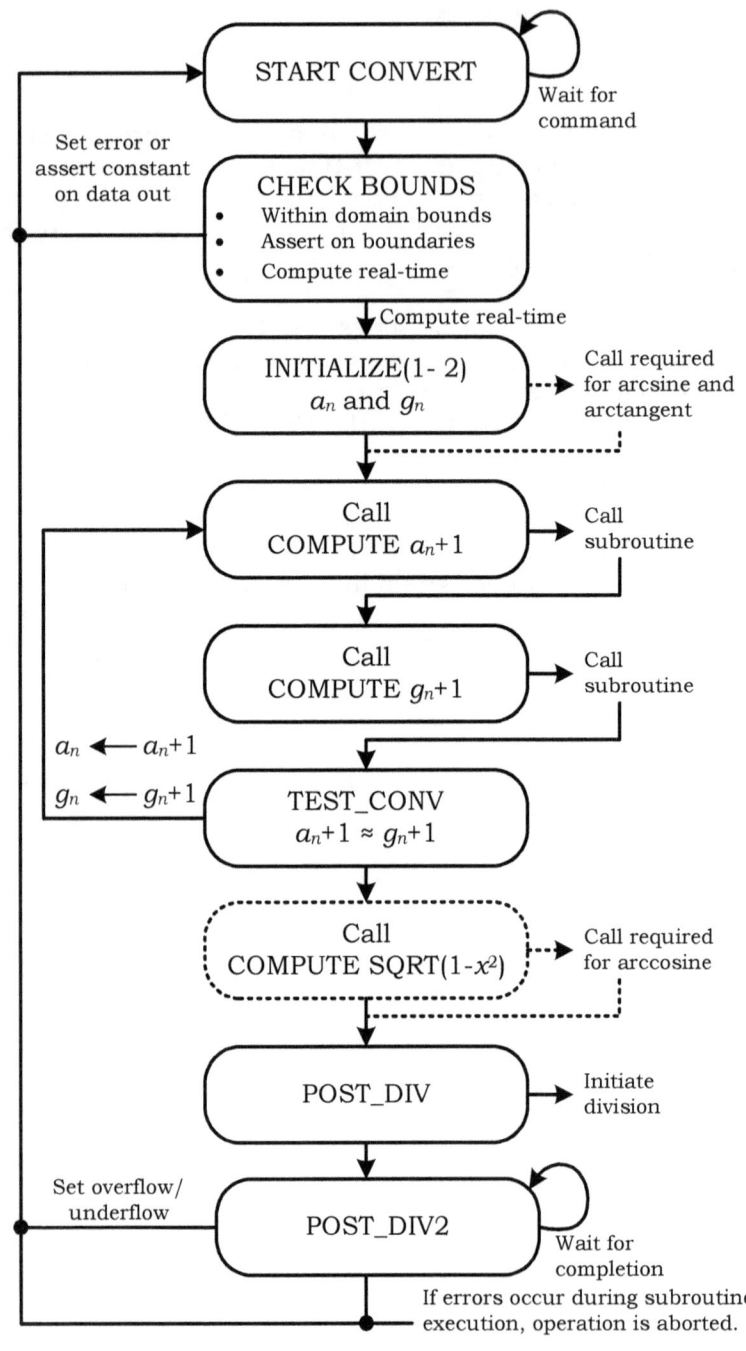

Figure 14

The table below shows which boundaries and which functions assert constant values. Even though the algorithm can handle the full domain range, rounding can interfere with an absolute zero when required, and with arccosine due to a divide by zero error when computing an x of -1.0.

Constant data asserted on domain boundaries

Functions	Domain Boundaries – versus – Output Data				
	> -∞	-1.0	0	+1.0	< +∞
Arcsine	N/A	computed	0	computed	N/A
Arccosine	N/A	+π	computed	0	N/A
Arctangent	computed	computed	0	computed	computed

The Input range and internal data formats are the same for arcsine and arccosine, but are different for arctangent. Arctangent requires a larger data input format. Output formats are the same for all three.

Input to output range

Functions	Input Range	Output Range
Arcsine	-1.0 to +1.0	$-\pi/2$ to $+\pi/2$
Arccosine	-1.0 to +1.0	$+\pi$ to 0
Arctangent	> -64 to < +64	> $-\pi/2$ to < $+\pi/2$

Note: *the input range for arctangent could be extended to a larger Qm.n. Compatibility was sought with the Table-Based and Taylor-Series designs.*

Several states call subroutines listed in the table below instead of repeating those operations elsewhere. This simplifies the design and supports the concept of reuse. All data objects are global and can be accessed by other states. A general rule is to limit the states involved in updating a data object but allow the reading of those objects by all states.

Subroutines used in design

Subroutine	Function	Destination	What design use
Compute a_{n+1}	$\frac{1}{2}(a_n + g_n)$	a_{n+1}	Use by all
Compute g_{n+1}	$\sqrt{((a_{n+1}) \bullet g_n)}$	g_{n+1}	Use by all
Compute sqrt($1+x^2$)	$\sqrt{(1-x^2)}$	root	For arcsine/arccosine
Compute sqrt($1-x^2$)	$\sqrt{(1+x^2)}$	root	For arctangent only
Compute sqrt()	$\sqrt{}$	root	Sub-subroutine

Some subroutines are slightly different from function to function. The latter is illustrated by the dotted line in Figure 14.

External Resources

All three designs require an external multiplier, divider, and square root function. Signal pins being available at the entity level of each module. Volumes 2, 3, and 5 of this book series provide usable arithmetic modules that can serve as these external resources. Overall performance is directly tied to their throughput latency (clock cycles).

Caveat: *Rounding is a function of the external modules used.*

Convergence

Borchardt-Gauss is referred to as an iterative mean requiring multiple cycles, iterations, in order to converge the arithmetic and geometric components. After each cycle a_n+1 is subtracted from g_n+1. The closer the difference is to zero, the better the convergence and the more accurate the result. In theory, after enough cycles, zero will be reached.

Zero is not necessary, only that the result be within an expected tolerance. Limiting the number of fractional bits that are evaluated reduces the number of cycles. All data formats default to 32 fractional bits in the fixed-point format (which is extremely accurate). A constant call CONV_FRAC_SIZ in the package file, which defaults to 20, specifies that the most significant 20-bits of the 32 fractional bits are used during the evaluation after subtracting a_n+1 from g_n+1.

The Log2(10) rule, which is ~3.322 bits per decimal number, results in having an equivalent decimal accuracy of 6 decimal places. When computing the arcsine function, 3 to 11 iterations are required to compute a given value across the entire input range.

Reducing the constant CONV_FRAC_SIZ to 10 from 20, results in 3 to 7 interations for convergence over the same input range. Consequently, accuracy is reduced by 1 to 2%. The number of iterations, the number of fractional bits used by operand, and performance of external modules, represent parameters that can be dialed in by the user.

Performance

Function	Targeted Device	Clock Speed
Sine	XC5VLX30	250MHz
Cosine	XC5VLX30	250MHz
Tangent	XC5VLX50(1)	229MHz

Note(1): *The logic resources required for arctangent can fit within a XC5VLX30, but the IOB count is too high because of the wider data format. External modules would normally be elsewhere in the same FPGA or ASIC device. The XC5VLX50 was selected because of the IOB requirement when using external modules.*

4.2.2 Borchardt-Gauss Package File

All Borchardt-Gauss functions require a short package file named *BG_RealTimePackage.vhd*.

Design Parameters

- The input data format for both arcsine and arccosine defaults to Q1.32, supporting an input range from -1.0 to +1.0. An extended format is provided for arctangent, Q6.32, supporting an input range from -63.9999999998 to +63.9999999998.
- Internal data paths for arcsine and arccosine default to Q1.32 for multiplication and square root, and Q2.32 for division. The extended format for arctangent are Q13.32 and Q6.32, respectively.
- Data out for all functions use Q2.32 for a range of $+/- (\pi/2)$ radians for arcsine and arctangent, and 0 to π for arccosine. Additional circuits in the arcsine and arctangent modules check upper quotient bits for overflow in addition to the overflow signal from the divider.

Note: *Although the number of fractional bits can be increased, all other data formats within the design must have the same number of fractional bits for the resize operator to work properly.*

Organization Within the Header

- Constants defining data formats for input, internal, and output data paths.

- Constants for comparing input domain boundaries for arcsine and arccosine, and output constants. The range for arctangent is dictated by the extended Qm.n input format.

Functions

- No functions

```
---------------------------------------------------------
--
--   BG_RealtimePackage.vhd (Borchardt-Gauss)
--
---------------------------------------------------------
library IEEE;
use IEEE.std_logic_1164.all;
use IEEE.numeric_std.all;
use IEEE.math_real.all;
use work.ConversionPackageV2.all;

--  package header
package BG_RealtimePackage is
    --
    --   Fixed-point data formats
    --
    -- standard data input for arcsine and arccosine, +/- 1.0, and arctangent
    -- +/- 63.9999999998. Having the same number of fractional bits for all
    -- data formats ensures the resize operator works properly.
    --
    -- standard input format  +/- 1.0 (arcsine/arccosine)
    constant DI_INT_SIZ: natural := 1; -- integer bits
    constant DI_FRAC_SIZ: natural := 32; -- fractional bits
    constant DI_SIZ: natural := 1 + DI_INT_SIZ + DI_FRAC_SIZ;
    -- standard divider format
    constant D_INT_SIZ: natural := 2;
    constant D_FRAC_SIZ: natural := DI_FRAC_SIZ;
    constant D_SIZ: natural := 1 + D_INT_SIZ + D_FRAC_SIZ;
    -- standard multiplier/square root format
    constant M_INT_SIZ: natural := 1;
    constant M_FRAC_SIZ: natural := DI_FRAC_SIZ;
    constant M_SIZ: natural := 1 + M_INT_SIZ + M_FRAC_SIZ;
    -- standard data output format (0 to pi radians)
    constant DO_INT_SIZ: natural := 2; -- integer bits
    constant DO_FRAC_SIZ: natural := DI_FRAC_SIZ; -- fractional bits
    constant DO_SIZ: natural := 1 + DO_INT_SIZ + DO_FRAC_SIZ;
```

```
-- extended input format +/- 63.9999999998 (arctangent)
constant XDI_INT_SIZ: natural := 6; -- integer bits
constant XDI_FRAC_SIZ: natural := DI_FRAC_SIZ; -- fractional bits
constant XDI_SIZ: natural := 1 + XDI_INT_SIZ + XDI_FRAC_SIZ;
-- extended divider format
constant XD_INT_SIZ: natural := 6;
constant XD_FRAC_SIZ: natural := DI_FRAC_SIZ;
constant XD_SIZ: natural := 1 + XD_INT_SIZ + XD_FRAC_SIZ;
-- extended multiplier/square root format
constant XM_INT_SIZ: natural := 13;
constant XM_FRAC_SIZ: natural := DI_FRAC_SIZ;
constant XM_SIZ: natural := 1 + XM_INT_SIZ + XM_FRAC_SIZ;
-- no extended output

-- constants standard input boundaries (arcsine and arccosine)
constant \+1.0\: signed(DI_SIZ-1 downto 0) :=
    RealToQmn(1.0,DI_INT_SIZ,DI_FRAC_SIZ,TZERO);
constant \-1.0\: signed(DI_SIZ-1 downto 0) :=
    RealToQmn(-1.0,DI_INT_SIZ,DI_FRAC_SIZ,TZERO);
-- constants for output quadrant boundaries
constant \0\: signed(DO_SIZ-1 downto 0) := (others=>'0');
constant \+pi/2\: signed(DO_SIZ-1 downto 0) :=
    RealToQmn(+1.5708,DO_INT_SIZ,DO_FRAC_SIZ,AZERO);
constant \-pi/2\: signed(DO_SIZ-1 downto 0) :=
    RealToQmn(-1.5708,DO_INT_SIZ,DO_FRAC_SIZ,AZERO);
constant \+pi\: signed(DO_SIZ-1 downto 0)  :=
    RealToQmn(+3.1416,DO_INT_SIZ,DO_FRAC_SIZ,AZERO);

-- number of fractional bits used for convergence testing (lesser significant bits -
-- are not included)
constant CONV_FRAC_SIZ: natural := 20;
    -- CONVERGENCE BITS/3.322 = equivalent decimal places

end;

--   package body
package body BG_RealtimePackage is

end;
```

4.2.3 Arcsine Function (Borchardt-Gauss)

The dashed line represents the results of the simulated module, the solid represents the ideal value.

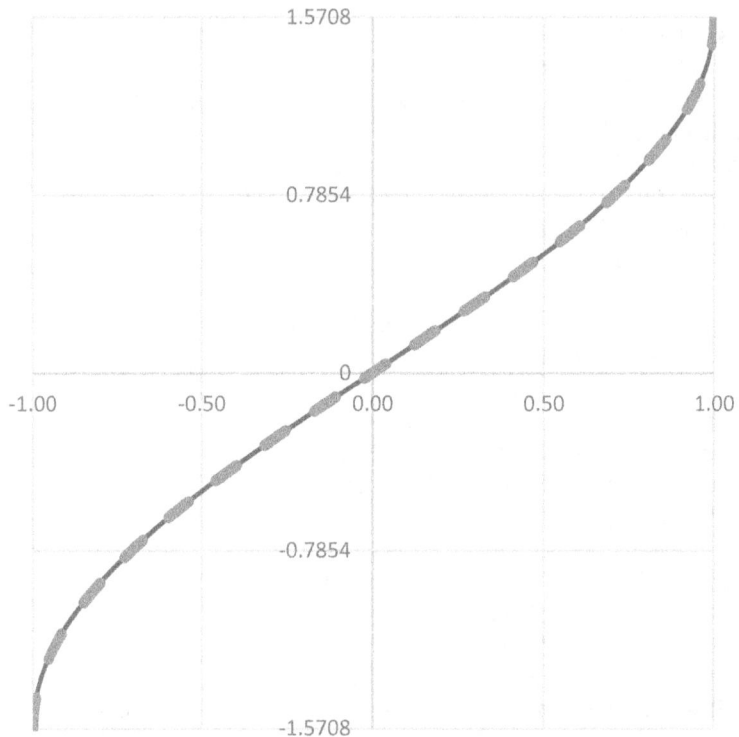

Recap of state machine design for Arcsine function (Borchardt-Gauss)

- Design support Q1.32 for an input range of +/- 1.0.
- Output format Q2.32 spanning +/- (π/2) radians. While the integer portion of the divider is 2 bits, extra circuits exist to detect an integer portion equaling or exceeding +/- 2 as an overflow.
- An *err* status bit indicates the input has exceeded +/-1.0. An *ovrflw* bit indicates an overflow occurred in one of the external modules, *udrflw* indicate an underflow. Either means the data is invalid.

4.2.3.1   Arcsine Function (source code)

```
-------------------------------------------------------------------------
--   ArcsineUsingBG.vhd (radians as input)(BG-Borchardt Gauss)
-------------------------------------------------------------------------
library IEEE;
use IEEE.std_logic_1164.all;
use IEEE.numeric_std.all;
use IEEE.math_real.all;
-- user packages
use work.ConversionPackageV2.all;
use work.BG_RealTimePackage.all;

entity ArcsineUsingBG is
--   Input and output fixed-point formats, as well as internal data paths,
--   are defined in BG_RealTimePackage.vhd
port
(
    clk,rst: in std_logic; -- system clock and reset (synchronous)
    -- inputs
    cnvrt: in std_logic; -- initiate arcsine function
    func_of_angle: in signed(DI_SIZ-1 downto 0);
    -- interface to external multiplier 1
    start1: out std_logic; -- start multiplication
    multiplier1: out signed(M_SIZ-1 downto 0);
    multiplicand1: out signed(M_SIZ-1 downto 0);
    cmplt1: in std_logic; -- multiplication complete
    ovrflw1: in std_logic; -- overflow error
    product1: in signed(M_SIZ-1 downto 0);
    -- interface to external square root function
    start2: out std_logic; -- start multiplication
    sqrt_arg2: out signed(M_SIZ-1 downto 0);
    cmplt2: in std_logic; -- square root complete
    ovrflw2: in std_logic; -- overflow error
    udrflow2: in std_logic; -- underflow error
    root2: in signed(M_SIZ-1 downto 0);
    -- interface to external divider
    start3: out std_logic; -- start division
    divisor3: out signed(D_SIZ-1 downto 0);
    dividend3: out signed(D_SIZ-1 downto 0);
    cmplt3: in std_logic; -- division complete
    ovrflw3: in std_logic; -- overflow error
    udrflow3: in std_logic; -- underflow error
```

```vhdl
    quotient3: in signed(D_SIZ-1 downto 0);
    -- outputs
    rdy: out std_logic; -- data ready
    err: out std_logic; -- error, input out of range
    ovrflw: out std_logic; -- multipliers or dividers
    udrflow: out std_logic; -- division underflow
    radians: out signed(DO_SIZ-1 downto 0)
);
end ArcsineUsingBG;

architecture RTL of ArcsineUsingBG is

--   declared constants
constant \0.5\: signed(M_SIZ-1 downto 0) :=
RealToQmn(0.5,M_INT_SIZ,M_FRAC_SIZ,TZERO);
constant \1.0\: signed(M_SIZ-1 downto 0) :=
RealToQmn(1.0,M_INT_SIZ,M_FRAC_SIZ,TZERO);
-- number of fractional bits are limited during convergence testing
constant ALIAS_SIZ: natural := 1 + M_INT_SIZ + CONV_FRAC_SIZ;

-- signals
signal x: signed(M_SIZ-1 downto 0) := (others=>'0');
signal an: signed(M_SIZ-1 downto 0) := (others=>'0');
signal \an+1\: signed(M_SIZ-1 downto 0) := (others=>'0');
signal gn: signed(M_SIZ-1 downto 0) := (others=>'0');
signal \gn+1\: signed(M_SIZ-1 downto 0) := (others=>'0');
signal sum: signed(M_SIZ-1 downto 0) := (others=>'0');
signal prod: signed(M_SIZ-1 downto 0) := (others=>'0');
signal root: signed(M_SIZ-1 downto 0) := (others=>'0');

-- alias for convergence checking
alias \an+1 test bits\ is \an+1\(M_SIZ-1 downto M_SIZ-ALIAS_SIZ);
alias \gn+1 test bits\ is \gn+1\(M_SIZ-1 downto M_SIZ-ALIAS_SIZ);
-- alias for local overflow detection
alias upper_quo_bits is quotient3(D_SIZ-1 downto DO_SIZ-2);

-- state machine enumeration list
type sm_def is
(
    RESET,
    START_CNVRT,
    CHECK_BOUNDS,
    INITIALIZE,INITIALIZE2,
    BG_ALGORITHM1,BG_ALGORITHM2,
```

```
        TEST_CONV,
        POST_DIV,POST_DIV2,
        OVERFLOW_ERROR,
        UNDERFLOW_ERROR,
        \compute an+1\,
        \compute an+1 (step2)\,
        \compute an+1 (step3)\,
        \compute gn+1\,
        \compute gn+1 (step2)\,
        \compute gn+1 (step3)\,
        \compute sqrt(1-x2)\,
        \compute sqrt(1-x2) (step2)\,
        \compute sqrt(1-x2) (step3)\,
        \compute sqrt(1-x2) (step4)\,
        \compute sqrt(sqrt_arg)\,
        \compute sqrt(sqrt_arg) (step2)\
);
signal state, ret_state, sav_state: sm_def := RESET;

begin

-------------------------------------------------
--  Arcsine Function state machine (using Radians)
-------------------------------------------------
process(rst,clk)
begin
    if(rst='1') then
        -- local data objects
        x <= (others=>'0');
        an <= (others=>'0');
        \an+1\ <= (others=>'0');
        gn <= (others=>'0');
        \gn+1\ <= (others=>'0');
        sum <= (others=>'0');
        prod <= (others=>'0');
        root <= (others=>'0');

        -- interface to external modules
        start1 <= '0';
        multiplier1 <= (others=>'0');
        multiplicand1 <= (others=>'0');
        start2 <= '0';
        sqrt_arg2 <= (others=>'0');
        start3 <= '0';
        divisor3 <= (others=>'0');
```

```vhdl
    dividend3 <= (others=>'0');

    -- outputs
    rdy <= '0';
    err <= '0';
    ovrflw <= '0';
    udrflow <= '0';
    radians <= (others=>'0');

    -- states
    state <= RESET;
    ret_state <= RESET;
    sav_state <= RESET;

elsif rising_edge(clk) then
    -- one clock signals
    start1 <= '0';
    start2 <= '0';
    start3 <= '0';


    --
    --   State Machine body
    --
    case state is
        -- reset state
        when RESET =>
        state <= START_CNVRT;
        -- wait for signal to start conversion
        when START_CNVRT =>
        if(cnvrt = '1') then
                rdy <= '0';err <= '0';
                ovrflw <= '0';udrflow <= '0';
                x <= resize(func_of_angle,x'length);
                state <= CHECK_BOUNDS;
            end if;
        -- check to be within +/-1, assert zero if zero
        when CHECK_BOUNDS =>
            if(x < \-1.0\ or x > \+1.0\) then
                rdy <= '1';err <= '1';
                state <= START_CNVRT;
            elsif(x = 0) then
                rdy <= '1';
                radians <= \0\;
                state <= START_CNVRT;
```

```
        else
            -- compute radians from x
            state <= INITIALIZE;
        end if;
-- initialize an and gn based on BG requirements
when INITIALIZE =>
    state <= \compute sqrt(1-x2)\;
    ret_state <= INITIALIZE2;
when INITIALIZE2 =>
    an <= root;
    gn <= RealToQmn(1.0,M_INT_SIZ,M_FRAC_SIZ,TZERO);
    state <= BG_ALGORITHM1;
--
--  Borchardt-Gauss Algorithm
--
when BG_ALGORITHM1 =>
    state <= \compute an+1\;
    ret_state <= BG_ALGORITHM2;
when BG_ALGORITHM2 =>
    state <= \compute gn+1\;
    ret_state <= TEST_CONV;
when TEST_CONV =>
    -- test for convergence
    if((\gn+1 test bits\ - \an+1 test bits\) = 0) then
        state <= POST_DIV;
    else
        an <= \an+1\;
        gn <= \gn+1\;
        state <= BG_ALGORITHM1;
    end if;
-- post process converged value
when POST_DIV =>
    -- start division
    start3 <= '1';
    dividend3 <= resize(x,dividend3'length);
    divisor3 <= resize(\an+1\,divisor3'length);
    state <= POST_DIV2;
when POST_DIV2 =>
    -- wait for division to complete
    if(cmplt3 = '1') then
        if(ovrflw3 = '1') then
            state <= OVERFLOW_ERROR;
        elsif(udrflow3 = '1') then
            state <= UNDERFLOW_ERROR;
```

```
                -- quotient output is same as data output. limit output
                -- to +/- pi/2 radians.
                elsif(upper_quo_bits = (upper_quo_bits'range => '0')) or
                     (upper_quo_bits = (upper_quo_bits'range => '1')) then

                    rdy <= '1';
                    radians <= resize(quotient3,radians'length);
                    state <= START_CNVRT;
                else
                    state <= OVERFLOW_ERROR;
                end if;
            end if;
        -- error handling
        when OVERFLOW_ERROR =>
            rdy <= '1';ovrflw <= '1';
            state <= START_CNVRT;
        when UNDERFLOW_ERROR =>
            rdy <= '1';udrflow <= '1';
            state <= START_CNVRT;


        ------------------
        --
        --   Subroutines
        --
        ------------------
        --
        -- compute an+1  and store
        --
        when \compute an+1\ =>
            -- add an and ag
            sum <= an + gn;
            state <= \compute an+1 (step2)\;
        when \compute an+1 (step2)\ =>
            -- multiply sum by 1/2
            start1 <= '1';
            multiplier1 <= \0.5\;
            multiplicand1  <= sum;
            state <= \compute an+1 (step3)\;
        when \compute an+1 (step3)\ =>
            -- wait for multiplication to complete
            if(cmplt1 = '1') then
                -- check for error
                if(ovrflw1 = '1') then
                    state <= OVERFLOW_ERROR;
```

```
        else
            \an+1\ <= product1;
            state <= ret_state;
        end if;
    end if;
--
--  compute gn+1 and store
--
when \compute gn+1\ =>
    -- multiply an+1 by gn
    start1 <= '1';
    multiplier1 <= gn;
    multiplicand1  <= \an+1\;
    sav_state <= ret_state;
    state <= \compute gn+1 (step2)\;
when \compute gn+1 (step2)\ =>
    -- wait for multiplication to complete
    if(cmplt1 = '1') then
        -- check for error
        if(ovrflw1 = '1') then
            state <= OVERFLOW_ERROR;
        else
            sqrt_arg2 <= product1;
            state <= \compute sqrt(sqrt_arg)\;
            ret_state <= \compute gn+1 (step3)\;
        end if;
    end if;
when \compute gn+1 (step3)\ =>
    -- store gn, then return
    \gn+1\ <= root;
    state <= sav_state;
--
--  compute sqrt(1-x2)
--
when \compute sqrt(1-x2)\ =>
    -- multiply x by x
    start1 <= '1';
    multiplier1 <= x;
    multiplicand1  <= x;
    state <= \compute sqrt(1-x2) (step2)\;
when \compute sqrt(1-x2) (step2)\ =>
    -- wait for multiplication to complete
    if(cmplt1 = '1') then
        -- check for error
```

```
                if(ovrflw1 = '1') then
                    state <= OVERFLOW_ERROR;
                else
                    prod <= product1;
                    state <= \compute sqrt(1-x2) (step3)\;
                end if;
            end if;
        when \compute sqrt(1-x2) (step3)\ =>
            -- subtract x2 from 1
            sum <= \1.0\ - prod;
            state <= \compute sqrt(1-x2) (step4)\;
        when \compute sqrt(1-x2) (step4)\ =>
            -- run square root then return
            sqrt_arg2 <= sum;
            -- return state set by caller
            state <= \compute sqrt(sqrt_arg)\;
        --
        --   compute square root of argument
        --
        when \compute sqrt(sqrt_arg)\ =>
            -- execute the sqrt function with whats in sqrt_arg
            start2 <= '1';
            state <= \compute sqrt(sqrt_arg) (step2)\;
        when \compute sqrt(sqrt_arg) (step2)\ =>
            -- wait for operation to comlete
            if(cmplt2 = '1') then
                -- check for error
                if(ovrflw2 = '1') then
                    state <= OVERFLOW_ERROR;
                elsif(udrflow2 = '1') then
                    state <= UNDERFLOW_ERROR;
                else
                    root <= root2;
                    state <= ret_state;
                end if;
            end if;
        when others =>
            state <= RESET;
        end case;
    end if;
end process;

end RTL;
```

4.2.4 Arccosine Function (Borchardt-Gauss)

The dashed line represent the results of the simulated module, the solid represents the ideal value.

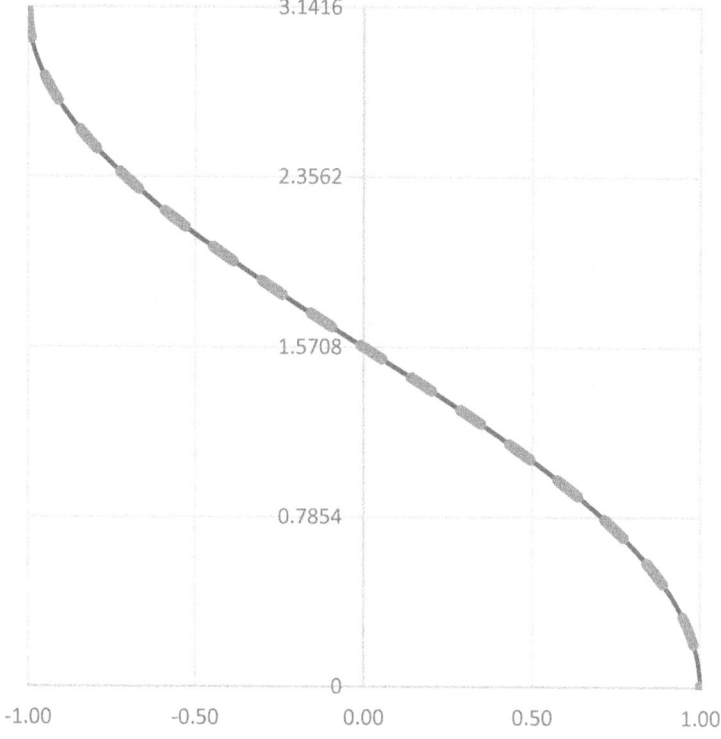

Recap on state machine design for Arccosine function (Borchardt-Gauss)

- Design support Q1.32 for an input range of +/- 1.0.
- Output format Q2.32 spanning 0 to π radians.
- An *err* status bit indicates the input has exceeded +/-1.0. An *ovrflw* bit indicates an overflow occurred in one of the external modules, *udrflw* indicate an underflow occurred. Either means the data is invalid.

4.2.4.1   Arccosine Function (source code)

Note: *Only differences are shown between the Borchardt-Gauss arcsine and the arccosine function because the source code is approximately identical.*

```
-----------------------------------------------------------------------
--  ArccosineUsingBG.vhd (radians as input)(BG-Borchardt Gauss)
-----------------------------------------------------------------------
.
entity ArccosineUsingBG is
.
end ArccosineUsingBG;

architecture RTL of ArccosineUsingBG is
.
-- state machine enumeration list
type sm_def is
(
        .
    INITIALIZE,
        .
);
.
begin
---------------------------------------------------
--  Arccosine Function state machine (using Radians)
---------------------------------------------------
process(rst,clk)
begin
    if(rst='1') then
            .
    elsif rising_edge(clk) then
        -- one clock signals
        start1 <= '0';
        start2 <= '0';
        start3 <= '0';

        --
        --   State Machine body
        --
        case state is
            .
```

```
-- check to be within +/-1, assert zero if zero
when CHECK_BOUNDS =>
    if(x < \-1.0\ or x > \+1.0\) then
        rdy <= '1';err <= '1';
        state <= START_CNVRT;
    elsif(x = \-1.0\) then
        rdy <= '1';
        radians <= \+pi\;
        state <= START_CNVRT;
    elsif(x = \+1.0\ )then
        rdy <= '1';
        radians <= \0\;
        state <= START_CNVRT;
    else
        -- compute radians from x
        state <= INITIALIZE;
    end if;
-- initialize an and gn based on BG requirements
when INITIALIZE =>
    an <= x;
    gn <= RealToQmn(1.0,M_INT_SIZ,M_FRAC_SIZ,TZERO);
    state <= BG_ALGORITHM1;
--
--   Borchardt-Gauss Algorithm
--
.
.
when TEST_CONV =>
    -- test for convergence
    if((\gn+1 test bits\ - \an+1 test bits\) = 0) then
        -- create dividend before divide
        state <= \compute sqrt(1-x2)\;
        ret_state <= POST_DIV;
    else
        an <= \an+1\;
        gn <= \gn+1\;
        state <= BG_ALGORITHM1;
    end if;
-- post process converged value
when POST_DIV =>
    -- start division
    start3 <= '1';
    dividend3 <= resize(root,dividend3'length);
    divisor3 <= resize(\an+1\,divisor3'length);
```

```vhdl
                    state <= POST_DIV2;
                when POST_DIV2 =>
                    -- wait for division to complete
                    if(cmplt3 = '1') then
                        if(ovrflw3 = '1') then
                            state <= OVERFLOW_ERROR;
                        elsif(udrflow3 = '1') then
                            state <= UNDERFLOW_ERROR;
                        else
                            rdy <= '1';
                            radians <= resize(quotient3,radians'length);
                            state <= START_CNVRT;
                        end if;
                    end if;

                  .

                  .
        end case;
    end if;
end process;

end RTL;
```

4.2.5 Arctangent Function (Borchardt-Gauss)

The dashed line represent the results of the simulated module, the solid represents the ideal value.

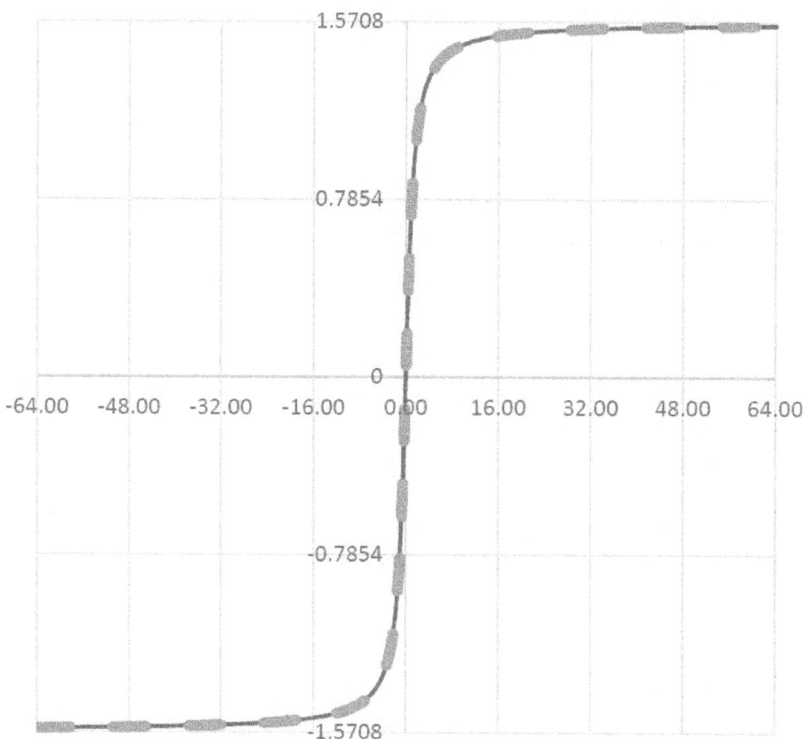

Recap on state machine design for Arctangent function (Borchardt-Gauss)

- Design support Q6.32 for an input range of +/- 63.9999999998.
- Output format Q2.32 spanning > (- π/2) radians to < (+ π/2) radians. While the integer portion of the divider is 6 bits, extra circuits exist to detect the integer portion equaling or exceeding +/- 2 as an overflow.
- The *ovrflw* bit indicates an overflow occurred in one of the external modules, *udrflw* indicate an underflow. Either means the data is invalid.

4.2.5.1   Arctangent Function (source code)

Note: *Only differences are shown between the Borchardt-Gauss arcsine and the arctangent function because the source code is approximately identical.*

```
--------------------------------------------------------------------------------
--  ArctangentUsingBG.vhd (radians as input)(BG-Borchardt Gauss)
--------------------------------------------------------------------------------
.
entity ArctangentUsingBG is
--  Input and output fixed-point formats, as well as internal data paths,
--  are defined in BG_RealTimePackage.vhd
port
(
    .
    -- inputs
    cnvrt: in std_logic; -- initiate arctangent function
    func_of_angle: in signed(XDI_SIZ-1 downto 0);
    -- interface to external multiplier 1
    .
    multiplier1: out signed(XM_SIZ-1 downto 0);
    multiplicand1: out signed(XM_SIZ-1 downto 0);
    .
    product1: in signed(XM_SIZ-1 downto 0);
    .
    sqrt_arg2: out signed(XM_SIZ-1 downto 0);
    .
    root2: in signed(XM_SIZ-1 downto 0);
    -- interface to external divider
    .
    divisor3: out signed(XD_SIZ-1 downto 0);
    dividend3: out signed(XD_SIZ-1 downto 0);
    .
    quotient3: in signed(XD_SIZ-1 downto 0);
    .
);
end ArctangentUsingBG;

architecture RTL of ArctangentUsingBG is

--  declared constants
```

```vhdl
constant \0.5\: signed(XM_SIZ-1 downto 0) :=
RealToQmn(0.5,XM_INT_SIZ,XM_FRAC_SIZ,TZERO);
constant \1.0\: signed(XM_SIZ-1 downto 0) :=
RealToQmn(1.0,XM_INT_SIZ,XM_FRAC_SIZ,TZERO);

-- number of fractional bits are limited during convergence testing
constant ALIAS_SIZ: natural := 1 + XM_INT_SIZ + CONV_FRAC_SIZ;

-- signals
signal x: signed(XM_SIZ-1 downto 0) := (others=>'0');
signal an: signed(XM_SIZ-1 downto 0) := (others=>'0');
signal \an+1\: signed(XM_SIZ-1 downto 0) := (others=>'0');
signal gn: signed(XM_SIZ-1 downto 0) := (others=>'0');
signal \gn+1\: signed(XM_SIZ-1 downto 0) := (others=>'0');
signal sum: signed(XM_SIZ-1 downto 0) := (others=>'0');
signal prod: signed(XM_SIZ-1 downto 0) := (others=>'0');
signal root: signed(XM_SIZ-1 downto 0) := (others=>'0');

-- alias for convergence checking
alias \an+1 test bits\ is \an+1\(XM_SIZ-1 downto XM_SIZ-ALIAS_SIZ);
alias \gn+1 test bits\ is \gn+1\(XM_SIZ-1 downto XM_SIZ-ALIAS_SIZ);
-- alias for overflow detection
alias upper_quo_bits is quotient3(XD_SIZ-1 downto DO_SIZ-2);

-- state machine enumeration list
type sm_def is
(
    .
    \compute sqrt(1+x2)\,
    \compute sqrt(1+x2) (step2)\,
    \compute sqrt(1+x2) (step3)\,
    \compute sqrt(1+x2) (step4)\,
    .
);
.
begin
-----------------------------------------------------
--   Arctangent Function state machine (using Radians)
-----------------------------------------------------
process(rst,clk)
begin
    .
    elsif rising_edge(clk) then
        -- one clock signals
```

```
start1 <= '0';
start2 <= '0';
start3 <= '0';


--
--   State Machine body
--
case state is

    .
    -- wait for signal to start conversion
    when START_CNVRT =>
        if(cnvrt = '1') then
            rdy <= '0';ovrflw <= '0';udrflow <= '0';

            x <= resize(func_of_angle,x'length);
            state <= CHECK_BOUNDS;
        end if;
    -- no restrictions on input, full Qm.n format
    when CHECK_BOUNDS =>
        if(x = 0) then
            rdy <= '1';
            radians <= \0\;
            state <= START_CNVRT;
        else
            -- compute radians from x
            state <= INITIALIZE;
        end if;
    -- initialize an and gn based on BG requirements
    when INITIALIZE =>
        state <= \compute sqrt(1+x2)\;
        ret_state <= INITIALIZE2;
    when INITIALIZE2 =>
        an <= RealToQmn(1.0,XM_INT_SIZ,XM_FRAC_SIZ,TZERO);
        gn <= root;
        state <= BG_ALGORITHM1;
    --
    --   Borchardt-Gauss Algorithm
    --
    .
    when POST_DIV2 =>
        -- wait for division to complete
        if(cmplt3 = '1') then
            if(ovrflw3 = '1') then
                state <= OVERFLOW_ERROR;
```

```
        elsif(udrflow3 = '1') then
            state <= UNDERFLOW_ERROR;
        -- quotient output is larger than data output. limit output
        -- to +/- pi/2 radians.
        elsif(upper_quo_bits = (upper_quo_bits'range => '0')) or
            (upper_quo_bits = (upper_quo_bits'range => '1')) then

            rdy <= '1';
            radians <= resize(quotient3,radians'length);
            state <= START_CNVRT;
        else
            state <= OVERFLOW_ERROR;
        end if;
    end if;


-----------------
--
--  Subroutines
--
-----------------

--
--  compute sqrt(1+x2)
--
when \compute sqrt(1+x2)\ =>
    -- multiply x by x
    start1 <= '1';
    multiplier1 <= x;
    multiplicand1  <= x;
    state <= \compute sqrt(1+x2) (step2)\;
when \compute sqrt(1+x2) (step2)\ =>
    -- wait for multiplication to complete
    if(cmplt1 = '1') then
        -- check for error
        if(ovrflw1 = '1') then
            state <= OVERFLOW_ERROR;
        else
            prod <= product1;
            state <= \compute sqrt(1+x2) (step3)\;
        end if;
    end if;
when \compute sqrt(1+x2) (step3)\ =>
    -- add x2 to 1
```

```vhdl
                sum <= \1.0\ + prod;
                state <= \compute sqrt(1+x2) (step4)\;
            when \compute sqrt(1+x2) (step4)\ =>
                -- run square root then return
                sqrt_arg2 <= sum;
                -- return state set by caller
                state <= \compute sqrt(sqrt_arg)\;


        end case;
    end if;
end process;

end RTL;
```

5 Addendum

The source code that follows supports conversions between real and fixed-point numbers, which exceed the 32-bit limits.

```vhdl
--
--   File: ConversionPackageV2.vhd
--
--   Changes from initial version:
--   1). Some synthesis tools will not allow parameters m or n
--   to be 0. Changes were made to support 0 values.
--   2). The RealToQmn function now allows for rounding.
--   3). Problem fixed on RealToQmn when integer > 30 bits
--
library IEEE;
use ieee.std_logic_1164.all;
use ieee.numeric_std.all;
use work.all;

--   package header
package ConversionPackageV2 is
    -- note: all numbers are processed as positive
    type r_def is
    (
        TZERO, -- truncate
        AZERO, -- round if any lower bits
        NEVEN -- .5 remainder and lsb even
    );

    function RealToQmn(r: real; m,n: integer; rnd: r_def) return signed;
    function QmnToReal(qmn: signed; m,n: integer) return real;

end;

--   package body
package body ConversionPackageV2 is


    --
    --   Convert real number to Qmn fixed-point number
    --
    --   m - number of integer bits excluding the sign (no limit)
    --   n - number of fractional bits (no limit)
    --   rnd - type of rounding operation
```

```vhdl
function RealToQmn(r: real; m,n: integer; rnd: r_def) return signed is
variable rx,tmp: real := 0.0;
variable int: unsigned(m-1 downto 0) := (others=>'0');
variable frac: unsigned(n-1 downto 0) := (others=>'0');
variable qmn: signed((m+n) downto 0) := (others=>'0');
begin
    -- convert to positive
    if(r < 0.0) then
        rx := (r * (-1.0));
    else
        rx := r;
    end if;
    -- compute integer portion
    if(m > 0)   then
        for i in m-1 downto 0 loop
            -- integer size limit
            if(i < 31) then
                -- subtract binary equivalent
                if((rx - real(2**i)) >= 0.0) then
                    rx := rx - real(2**i);
                    int(i) := '1';
                end if;
            -- multiply up limit by 2
            else
                tmp := real(2**30);
                --for j in (m-1) downto 31 loop
                for j in i downto 31 loop
                    tmp := tmp * 2.0;
                end loop;
                -- subtract binary equivalent
                if((rx - tmp) >= 0.0) then
                    rx := rx - tmp;
                    int(i) := '1';
                end if;
            end if;
        end loop;
    end if;
    -- compute fractional portion (remaining in rx)
    if(n > 0) then
        for i in n-1 downto 0 loop
            -- integer size limit
            if((n-i) < 31) then
                -- subtract reciprocal
                tmp := (1.0/real(2**(n-i)));
```

```
            if((rx - tmp) >= 0.0) then
                rx := rx - tmp;
                frac(i) := '1';
            end if;
        -- divide down limit by 2
        else
            tmp := 1.0/real(2**30);
            for j in 31 to (n-i) loop
                tmp := tmp / 2.0;
            end loop;
            -- subtract reciprocal
            if((rx - tmp) >= 0.0) then
                rx := rx - tmp;
                frac(i) := '1';
            end if;
        end if;
    end loop;
    -- adjust remainder for rounding
    for i in 1 to n loop
        rx := rx * 2.0;
    end loop;
end if;


-- construct final fixed-point number
if(m = 0 and n = 0) then
    report "m and n parameters cannot both be zero!" severity failure;
elsif(m = 0) then
    qmn := signed('0'&frac);
elsif(n = 0) then
    qmn := signed('0'&int);
else
    qmn := signed('0'&int&frac);
end if;


-- round according to parameter
if(rnd = AZERO and rx > 0.0) then
    qmn := qmn + 1;
elsif(rnd = NEVEN and (rx > 0.5 or (rx = 0.5 and qmn(qmn'low) = '1'))) then
    qmn := qmn + 1;
else
    null; -- truncate
end if;


-- convert to negative if needed
```

```vhdl
    if(r < 0.0) then
        qmn := (not qmn) + 1;
    end if;

    return(qmn);

end function;
--
--   Convert Qmn fixed-point number to real
--
--   m - number of integer bits excluding the sign (no limit)
--   n - number of fractional bits (no limit)
--
function QmnToReal(qmn: signed; m,n: integer) return real is
variable qmnx: signed(qmn'high downto qmn'low) := (others=>'0');
variable r,tmp: real := 0.0;
begin
    -- convert to positive number
    if(qmn(qmn'high) = '1') then
        qmnx := (not qmn) + 1;
    else
        qmnx := qmn;
    end if;
    -- compute integer portion
    if(m > 0) then
        -- add corresponding power of 2
        for i in qmnx'high-m to qmnx'high-1 loop
            if(qmnx(i) = '1') then
                -- integer size limit
                if((i-n) < 31) then
                    r := r + real(2**(i-n));
                -- multiply up limit by 2
                else
                    tmp := real(2**30);
                    for j in (i-n) downto 31 loop
                        tmp := tmp * 2.0;
                    end loop;
                    r := r + tmp;
                end if;
            end if;
        end loop;
    end if;
    -- compute fractional portion
    if(n > 0) then
```

```
        -- add corresponding power of two reciprocal
        for i in qmnx'high-1-m downto qmnx'low loop
            if(qmnx(i) = '1') then
                -- integer size limit
                if((n-i) < 31) then
                    r := r + (1.0/real(2**(n-i)));
                -- divide down limit by 2
                else
                    tmp := 1.0/real(2**30);
                    for j in 31 to (n-i) loop
                        tmp := tmp / 2.0;
                    end loop;
                    r := r + tmp;
                end if;
            end if;
        end loop;
    end if;
    -- restore sign
    if(qmn(qmn'high) = '1') then
        r := (r * (-1.0));
    end if;

    return(r);

end function;

end;
```

www.ingramcontent.com/pod-product-compliance
Lightning Source LLC
Chambersburg PA
CBHW051908170526
45168CB00001B/289